全栈开发

Spring Cloud
微服务架构实战

周喜平 / 主编

人民邮电出版社

北京

图书在版编目（CIP）数据

Spring Cloud微服务架构实战 / 周喜平主编. -- 北京：人民邮电出版社，2022.9
ISBN 978-7-115-59769-4

Ⅰ．①S… Ⅱ．①周… Ⅲ．①互联网络—网络服务器 Ⅳ．①TP368.5

中国版本图书馆CIP数据核字(2022)第135305号

内 容 提 要

本书以实战化训练为宗旨，用详尽且经典的案例讲述 Spring Cloud 的项目搭建方法和常用技术。每个案例都配有详细讲解和代码，可以帮助读者快速掌握书中的各个知识点。

本书首先介绍系统架构的发展历史、常见的微服务架构、Spring Boot 和 Spring Cloud 的关系；然后介绍 Spring Cloud 开发环境的搭建，如 JDK、IntelliJ IDEA、Maven 的安装和配置；继而介绍微服务环境的创建、微服务项目案例的搭建和微服务的调用；最后介绍常用的微服务技术，如 Eureka、Ribbon、Feign、Hystrix、Spring Cloud Gateway、Spring Cloud Stream 和 Spring Cloud Config。

本书适合具备 Java 基础的开发人员、对微服务架构和 Spring Cloud 感兴趣的读者、想要了解 Spring 或 Spring Cloud 的开发人员阅读。对尝试选择或实施微服务架构的团队来说，本书具有较高的参考价值。

◆ 主　　编　周喜平
　 责任编辑　张天怡
　 责任印制　陈　犇

◆ 人民邮电出版社出版发行　　北京市丰台区成寿寺路11号
　 邮编　100164　　电子邮件　315@ptpress.com.cn
　 网址　https://www.ptpress.com.cn
　 三河市中晟雅豪印务有限公司印刷

◆ 开本：787×1092　1/16
　 印张：18.5　　　　　　　　　　2022年9月第1版
　 字数：473千字　　　　　　　　 2022年9月河北第1次印刷

定价：69.80元

读者服务热线：(010)81055410　印装质量热线：(010)81055316
反盗版热线：(010)81055315
广告经营许可证：京东市监广登字20170147号

前言
PREFACE

为什么要写这本书

在互联网时代，互联网产品的一大特点就是需要快速发布新功能，支持高并发和大数据。

传统的架构已经慢慢不能支撑互联网业务的发展，这时候微服务架构顺势而出。

一开始国内很多公司都是基于开源的 Dubbo 框架来构建微服务的，由于某些原因，Dubbo 已经很久没进行维护了，不过 2018 年又重新开始维护。反观 Spring Cloud，其在国外发展得很好，但几年前，其在国内还鲜为人知。不过从 2017 年开始，Spring Cloud 在国内的普及度越来越高，很多中小型互联网公司都开始"拥抱"Spring Cloud。

Spring Cloud 提供了一整套微服务的解决方案，基于 Spring Boot 可实现快速集成，且开发效率很高，因而可以说它为中小型互联网公司带来了"福音"。而且 Spring Cloud 发布新功能的频率非常高，目前仅大版本就有很多个，同时它还有庞大的社区支持，照这样的发展势头，我相信未来几年国内 Spring Cloud 的发展将蒸蒸日上。

我一直在使用 Spring Boot、Spring Data 等框架来进行开发工作。作为一名 Spring 系列的忠实粉丝，我自然希望能够有更多的开发者参与进来，于是自己坚持写 Spring Cloud 相关的文章，并且将文章涉及的代码整理后放在 GitHub 上分享。这使我得到了很多开发者朋友的关注，他们向我咨询一些微服务方面的问题，我也会自己研究和解决一些问题，然后通过文章的形式将解决方法分享给各位开发者朋友。我写本书的一个目的是想进一步推动 Spring Cloud 在国内的使用和发展，同时分享自己在微服务领域取得的一些小小的经验。

另外，本书在创作过程中，得到了河南省民办普通高等学校专业建设资助项目的基金支持，是郑州西亚斯学院的专业建设的重要成果之一。

本书特色

- **入门级的讲解**

无论读者是否从事计算机相关行业，是否接触过 Spring Cloud，是否使用 Spring Cloud 开发过项目，都能从本书中获益。

● **大量实用、专业的范例和项目**

本书结合开发者实际工作中的范例，逐一讲解 Spring Cloud 的相关知识和技术，帮助读者在实战中掌握知识，轻松拥有项目经验。

● **大量电子资源**

赠送大量电子资源，包括本书范例的素材文件和结果文件、本书教学 PPT 等。

本书内容

本书主要讲解的是如何建立与微服务相关的实战体系。第 1 章介绍认识微服务，可帮助读者了解微服务以及 Spring Cloud 的概念。第 2 章介绍准备开发环境，可帮助读者正确快速地搭建开发环境。第 3 章介绍贯穿案例，将电商项目实战案例呈现给读者，并引出服务调用，引导读者思考并了解微服务架构的结构体系。第 4~10 章介绍微服务架构中常用的技术并配合实战项目进行讲解。

创作团队

本书由周喜平主编，董丽莎、郝玉峰任副主编，王晓宇、邹佳丽参编。其中，第 7 章、第 9 章及第 10 章由郑州西亚斯学院的周喜平老师编著，第 2 章、第 5 章、第 8 章由郑州西亚斯学院的董丽莎老师编著，第 1 章和第 4 章由郑州西亚斯学院的郝玉峰老师编著，第 3 章由郑州西亚斯学院的邹佳丽老师编著，第 6 章由郑州西亚斯学院的王晓宇老师编著。

在本书的编写过程中，我们竭尽所能地将最好的讲解呈现给读者，但书中难免有疏漏和不妥之处，敬请广大读者不吝指正。若读者在阅读本书时遇到困难或产生疑问，或有任何建议，可发送邮件至 zhangtianyi@ptpress.com.cn。

编者

2022 年 8 月

001	第 1 章	认识微服务
002	1.1	系统架构的发展历史
002	1.1.1	单体架构
003	1.1.2	垂直架构
003	1.1.3	分布式架构
004	1.1.4	面向服务的架构
004	1.1.5	微服务架构
005	1.2	常见的微服务架构
006	1.2.1	Spring Cloud
008	1.2.2	Dubbo
009	1.2.3	Dropwizard
009	1.2.4	Cricket
009	1.2.5	Jersey
009	1.2.6	Play
009	1.3	Spring Boot 和 Spring Cloud 的关系
010	1.3.1	认识 Spring Boot
011	1.3.2	Spring Boot 整合 Spring Cloud
013	第 2 章	准备开发环境
014	2.1	Java 开发环境 JDK
014	2.1.1	下载 JDK
016	2.1.2	安装 JDK
018	2.1.3	配置 Java 环境变量
019	2.2	开发工具 IntelliJ IDEA
019	2.2.1	下载 IntelliJ IDEA
020	2.2.2	安装 IntelliJ IDEA
026	2.3	项目管理工具 Maven
027	2.3.1	下载 Maven
028	2.3.2	安装 Maven
028	2.3.3	配置 Maven 环境变量
029	2.3.4	配置 Maven 本地仓库和下载源
030	2.3.5	与 IntelliJ IDEA 集成
031	第 3 章	贯穿案例
032	3.1	数据库
033	3.2	创建工程

目 录
CONTENTS

033	3.2.1 创建父工程
035	3.2.2 创建子工程——用户微服务
040	3.2.3 创建子工程——商品微服务
045	3.2.4 创建子工程——订单微服务

- 049 3.3 使用 Postman 测试微服务
- 049   3.3.1 测试新增
- 050   3.3.2 测试查询全部
- 050   3.3.3 测试根据 id 查询单个
- 051   3.3.4 测试修改
- 052   3.3.5 测试删除
- 053 3.4 调用微服务
- 053   3.4.1 介绍 RestTemplate 类
- 053   3.4.2 使用 RestTemplate 调用微服务
- 054   3.4.3 分析硬编码存在的问题

第 4 章 Eureka 服务注册和发现

- 055
- 056 4.1 认识 Eureka
- 056   4.1.1 服务注册和服务发现
- 057   4.1.2 注册中心
- 059   4.1.3 Eureka 框架的原理
- 060 4.2 使用 Eureka
- 060   4.2.1 搭建 Eureka 注册中心
- 063   4.2.2 将服务注册到 Eureka 注册中心
- 064   4.2.3 使用 Eureka 的元数据完成服务调用
- 065 4.3 Eureka 服务端高可用集群
- 066   4.3.1 搭建 Eureka 服务端高可用集群
- 068   4.3.2 将服务注册到 Eureka 服务端集群
- 070 4.4 Eureka 常见问题
- 070   4.4.1 服务注册慢
- 070   4.4.2 服务节点剔除问题
- 071   4.4.3 监控页面显示 IP 地址信息
- 071 4.5 Eureka 源码解析
- 072   4.5.1 服务注册表
- 073   4.5.2 服务注册
- 075   4.5.3 接收服务心跳

076	4.5.4	服务剔除
078	4.5.5	服务下线
080	4.5.6	集群同步
084	4.5.7	获取注册表中服务实例的信息

089　第 5 章　基于 Ribbon 服务调用

090	5.1	认识 Ribbon
090	5.1.1	微服务之间的交互
091	5.1.2	Ribbon 的两个主要作用
091	5.1.3	客户端的负载均衡
093	5.2	基于 Ribbon 实现负载均衡调用
093	5.2.1	坐标依赖
094	5.2.2	工程改造
097	5.2.3	代码测试
098	5.3	Ribbon 源码解析
098	5.3.1	配置和实例初始化
100	5.3.2	负载均衡器
102	5.3.3	ILoadBalancer 的实现
105	5.3.4	负载均衡策略实现

115　第 6 章　基于 Feign 服务调用

116	6.1	认识 Feign
116	6.1.1	Java 项目中接口的调用方式
117	6.1.2	Feign 和 Ribbon 的关系
117	6.2	使用 Feign 实现服务调用
117	6.2.1	坐标依赖
117	6.2.2	工程改造
119	6.2.3	代码测试
120	6.3	Feign 自定义配置和使用
120	6.3.1	日志配置
122	6.3.2	超时时间配置
122	6.3.3	客户端组件配置
123	6.3.4	压缩配置
124	6.3.5	使用配置文件自定义 Feign 的配置
125	6.4	源码分析
125	6.4.1	核心组件与概念

126	6.4.2 动态注册 BeanDefinition
133	6.4.3 实例初始化
136	6.4.4 函数调用和网络请求

143	**第 7 章　Hystrix 服务熔断**
144	7.1　认识 Hystrix
144	7.1.1　雪崩效应
145	7.1.2　线程隔离
146	7.1.3　服务熔断
147	7.2　使用 REST 实现服务熔断
147	7.2.1　坐标依赖
147	7.2.2　工程改造
150	7.2.3　代码测试
150	7.3　使用 Feign 实现服务熔断
151	7.3.1　坐标依赖
151	7.3.2　工程改造
153	7.3.3　代码测试
153	7.4　使用 Hystrix 实现监控
154	7.4.1　使用 Hystrix Dashboard 查看监控数据
157	7.4.2　使用 Hystrix Turbine 聚合监控数据
159	7.4.3　断路器的状态
162	7.4.4　断路器的隔离策略
163	7.5　源码分析
164	7.5.1　封装 HystrixCommand
169	7.5.2　断路器逻辑

175	**第 8 章　Spring Cloud Gateway 服务网关**
176	8.1　认识 Spring Cloud Gateway
177	8.1.1　微服务网关概述
178	8.1.2　微服务网关工作流程
178	8.2　实现服务网关
179	8.2.1　创建子工程——服务网关
179	8.2.2　坐标依赖
179	8.2.3　工程改造

182		8.2.4 代码测试
183	8.3	路由规则
183		8.3.1 路由规则概述
188		8.3.2 动态路由
189		8.3.3 重写转发路径
191	8.4	过滤器
191		8.4.1 过滤器基础
192		8.4.2 局部过滤器
194		8.4.3 全局过滤器
196	8.5	网关限流
197		8.5.1 常见的限流算法
197		8.5.2 基于过滤器的限流
201		8.5.3 基于 Sentinel 的限流
205	8.6	源码解析
206		8.6.1 初始化配置
207		8.6.2 网关处理器
209		8.6.3 路由定义定位器
211		8.6.4 路由定位器
211		8.6.5 路由断言
212		8.6.6 网关过滤器
213		8.6.7 全局过滤器
213		8.6.8 API 端点

215	**第 9 章**	**Spring Cloud Stream 消息驱动**
216	9.1	认识 Spring Cloud Stream
216		9.1.1 消息队列
218		9.1.2 绑定器
219		9.1.3 发布订阅模式
220	9.2	实现消息驱动
220		9.2.1 安装 RabbitMQ
224		9.2.2 消息生产者
226		9.2.3 消息消费者
229		9.2.4 自定义消息通道
231	9.3	消费者组
232		9.3.1 工程改造
234		9.3.2 代码测试

234	9.4	消费分区
235		9.4.1 工程改造
237		9.4.2 代码测试
238	9.5	源码解析
239		9.5.1 动态注册 BeanDefinition
241		9.5.2 消息发送的流程
243		9.5.3 @StreamListener 注解的处理

249	**第 10 章**	**Spring Cloud Config 分布式配置中心**
250	10.1	认识 Spring Cloud Config
251		10.1.1 配置中心概述
251		10.1.2 其他配置中心
251	10.2	实现配置中心
251		10.2.1 配置管理
254		10.2.2 服务端
257		10.2.3 客户端
259		10.2.4 配置刷新
264	10.3	服务总线
264		10.3.1 消息代理
265		10.3.2 工程改造
270	10.4	源码解析
271		10.4.1 配置服务器
280		10.4.2 配置客户端

第 1 章

认识微服务

微服务架构是近两年在软件架构领域出现的一个新名词。虽然诞生时间不长,但其在各种演讲、文章、图书中出现的频率已经让很多人意识到它给软件领域所带来的影响。

到底什么是微服务架构呢?当我们谈论微服务时,它代表一种什么样的含义?它与传统的面向服务的体系结构(或称架构)(Service-Oriented Architecture,SOA)有什么异同点?本章,我们将揭开微服务的神秘面纱。

本章的主要内容如下。

1. 系统架构的发展历史。
2. 常见的微服务架构。
3. Spring Boot 和 Spring Cloud 的关系。

1.1 系统架构的发展历史

随着互联网的发展、网站应用规模的不断扩大、上网需求的激增，互联网技术上的压力不断增加，系统架构也因此不断地演进、升级、迭代，从单体架构，到垂直架构，到分布式架构，再到 SOA，以及现在火热的微服务架构。

很多软件都是基于最新架构开发的，很少有人会去了解以前的系统架构，但对于想往架构师方向发展的开发者来说，了解系统架构的发展历史和每个阶段的系统架构，是很有必要的，这同样能够帮助开发者更好地设计以及演进系统架构。

1.1.1 单体架构

早期，一般的公司在开发 Java Web 程序时，大都使用 Struts 2、Spring 和 Hibernate 等技术框架，每一个项目都会发布一个单体应用。例如开发一个进销存系统，会开发一个 WAR 包部署到 Tomcat 中，每次需要开发新的模块或添加新的功能时，都会在原来的基础上不断地添加。若干次后，这个 WAR 包会不断地膨胀，程序员在进行调试时，服务器可能需要很长时间才能启动，维护这个系统的效率极为低下。图 1-1 所示为电商系统的单体架构，涵盖了商品管理、订单管理、用户管理等模块。

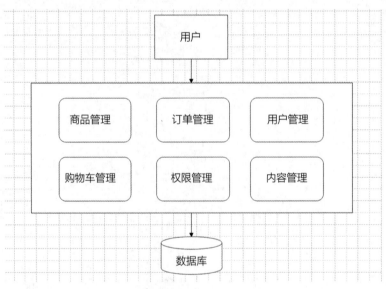

图1-1 单体架构

单体架构的优点和缺点如下。

优点：开发速度快，维护成本低，适用于并发要求较低的系统。

缺点：代码耦合度高，后期维护困难，无法根据不同模块进行针对性优化，无法水平扩展，单点容错率低，并发能力差。

1.1.2 垂直架构

当访问量逐渐增大，单体架构应用无法满足需求时，为了满足更高的并发和业务需求可以根据业务功能对系统进行拆分，以提高访问效率。电商系统的垂直架构如图1-2所示。

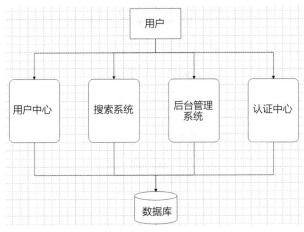

图1-2 垂直架构

垂直架构的优点和缺点如下。

优点：对系统进行拆分，实现了流量分担，解决了并发问题，可以针对不同模块进行优化，方便水平扩展、负载均衡，容错率提高。

缺点：系统间相互独立，会有很多重复的开发工作，影响开发效率。

1.1.3 分布式架构

当垂直应用越来越多时，应用之间的交互不可避免，可将核心业务抽取出来作为独立的服务，逐渐形成稳定的服务中心，使前端应用能更快速地响应多变的市场需求。电商系统的分布式架构如图1-3所示。

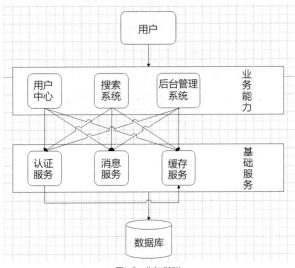

图1-3 分布式架构

分布式架构的优点和缺点如下。

优点：将基础服务进行了抽取，系统间相互调用，提高了代码复用率和开发效率。

缺点：系统间耦合度变高，调用关系错综复杂，难以维护。

1.1.4 面向服务的架构

面向服务的架构（SOA）是一种设计方法，其中包含多个服务，服务之间通过相互依赖最终提供一系列的功能。一个服务通常以独立的形式存在于操作系统进程中，各个服务之间通过网络调用。SOA示例如图1-4所示。

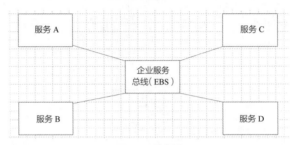

图1-4 SOA示例

企业服务总线（Enterprise Service Bus，ESB）简单来说就是管道，用来连接各个服务节点。为了集成不同系统、不同协议的服务，ESB做了消息的转化解释和路由工作，让不同的服务互联互通。

SOA的优点和缺点如下。

优点：抽取公共的功能为服务，提高了开发效率；对不同的服务进行集群化部署，缓解了系统压力；减少了系统耦合。

缺点：每个供应商提供的ESB产品有差异，自身实现较为复杂，应用服务粒度较大；ESB整合了所有服务、协议和数据转换，使得运维、测试、部署困难；所有服务都通过一条通路通信，直接降低了通信速率。

1.1.5 微服务架构

微服务架构使用一套小服务来开发单个应用，每个服务基于单一业务功能构建，运行在自己的进程中，使用轻量级通信机制，通常采用HTTP RESTful API（RESTful API是利用HTTP请求访问或使用数据的应用程序接口），能够通过自动化部署机制来独立部署。这些服务可以使用不同的编程语言实现，以适应不同的数据存储技术，并保持最低限度的集中式管理。微服务架构示例如图1-5所示。

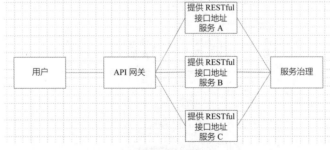

图1-5 微服务架构示例

网关（Gateway）通常是一个服务器，是系统的唯一入口，为每个客户端提供一个定制的 API。API 网关的核心是所有的客户端和服务器都通过统一的网关接入微服务，在网关实现所有的非业务功能。它还可以具有其他职责，如身份验证、监控、负载均衡、缓存、请求分片与管理、静态响应处理等。通常，网关提供 RESTful HTTP 的方式访问服务，而服务器通过注册中心进行服务注册和管理。

微服务的特点如下。

（1）单一职责：微服务中每一个服务都对应唯一的业务功能，做到单一职责。

（2）微：微服务的服务拆分粒度很小，例如用户管理就可以作为一个服务。每个服务虽小，但"五脏俱全"。

（3）面向服务：面向服务是指每个服务都要对外暴露 RESTful 接口 API，不关心服务的技术实现，做到与平台和语言无关，也不限定用什么技术实现，只要提供 REST 的接口即可。

（4）自治：自治是指服务间互相独立、互不干扰、耦合度低。团队独立是指每个服务都有一个独立的开发团队，人数不能过多；技术独立是指因为是面向服务，提供 REST 接口，使用什么技术没有别人干涉；前后端分离是指采用前后端分离开发，提供统一 REST 接口，后端不用再为 PC、移动端开发不同的接口；数据库分离是指每个服务都使用自己的数据源，部署独立，服务间虽然有调用，但能做到一个服务重启不影响其他服务，有利于持续集成和持续交付。每个服务都是独立的组件，可复用、可替换和易维护。

微服务架构与 SOA 都是对系统进行拆分；微服务架构基于 SOA 思想，把微服务当作去除了 ESB 的 SOA。ESB 是 SOA 中的中心总线，设计图形应该是星形的，而微服务架构是去中心化的分布式软件架构。两者比较类似，但其实也有一些差别，如表 1-1 所示。

表 1-1　SOA 与微服务架构对比

对比项	SOA	微服务架构
组件大小	大块业务逻辑	单独任务或小块业务逻辑
耦合	通常为松耦合	总是松耦合
公司架构	任何类型	小型、专注于功能交叉团队
管理	着重中央管理	着重分散管理
目标	确保应用能够交互操作	执行新功能、快速拓展开发团队

微服务架构的优点和缺点如下。

优点：通过服务的原子化拆分，以及微服务的独立打包、部署和升级，小团队的交付周期将缩短，运维成本也将大幅度下降；微服务遵循单一原则，微服务之间采用 RESTful 等轻量协议传输。

缺点：微服务过多，服务治理成本高，不利于系统维护；分布式系统开发的技术成本高（容错、分布式事务等）。

1.2　常见的微服务架构

随着程序规模的扩大以及复杂性的增加，越来越多的 Java 程序员选择使用微服务进行项目开发。微服务的出现有助于开发人员用更低的成本和更少的错误来开发程序，因此成为 Java 开发人员需要掌握的最重要的技术之一。下面介绍几种 Java 微服务架构。

1.2.1 Spring Cloud

Spring Cloud 是 Spring 旗下的项目之一，Spring 擅长的就是集成，把世界上的好框架"拿"过来，集成到自己的项目中。Spring Cloud 的官方简介如图 1-6 所示。

图1-6　Spring Cloud官方简介

图 1-6 中"OVERVIEW"的相关描述翻译如下。

Spring Cloud 为开发人员提供了工具来快速构建分布式系统中的一些常见模式（如配置管理、服务发现、断路器、智能路由、微代理、控制总线、一次性令牌、全局锁、领导选举、分布式会话、集群状态）。分布式系统的协调导致了"样板模式"，开发人员使用 Spring Cloud 可以快速实现这些模式的服务和应用程序。它们在任何分布式环境下都能很好地工作，包括开发者自己的笔记本计算机、裸机数据中心和云计算等托管平台。

Spring Cloud 是一个基于 Spring Boot 实现的微服务架构开发工具。Spring Cloud 有丰富的子组件，其架构如图 1-7 所示。

（1）Spring Cloud Config：配置和管理开发工具包，可以把配置放到远程服务器，目前支持本地存储，Git 存储以及 Subversion 存储。

（2）Spring Cloud Bus：事件、消息总线，用于在集群（如配置变化事件）中传播状态变化信息，可与 Spring Cloud Config 联合实现热部署。

（3）Spring Cloud Netflix：针对多种 Netflix 组件提供的开发工具包，包括 Eureka、Hystrix、Zuul、Archaius 等。

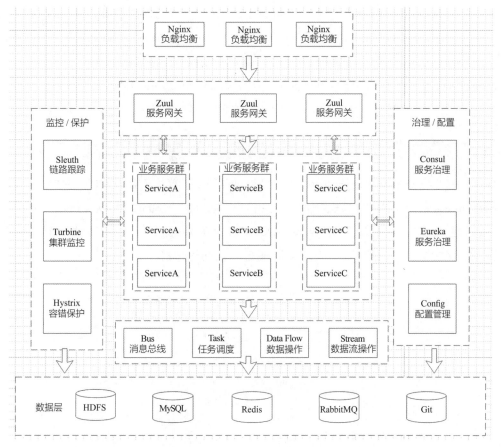

图1-7　Spring Cloud架构

（4）Spring Cloud Eureka：云端负载均衡，一个基于 REST 的服务，用于定位服务，以实现云端的负载均衡和中间层服务器的故障转移。

（5）Spring Cloud Hystrix：容错管理工具，旨在通过控制服务和第三方库的节点，对延迟和故障提供更强大的容错能力。

（6）Spring Cloud Zuul：边缘服务工具，提供动态路由、监控、弹性、安全等的边缘服务。

（7）Spring Cloud Archaius：配置管理 API，包含一系列配置管理 API，提供动态类型化属性、线程安全配置操作、轮询框架、回调机制等功能。

（8）Spring Cloud for Cloud Foundry：通过 OAuth 2.0 协议绑定服务到 Cloud Foundry，Cloud Foundry 是 VMware 推出的开源平台即服务（Platform as a Service，PaaS）云平台。

（9）Spring Cloud Sleuth：日志收集工具包，封装了 Dapper、Zipkin 和 HTrace 操作。

（10）Spring Cloud Data Flow：大数据操作工具，通过命令行方式操作数据流。

（11）Spring Cloud Security：安全工具包，为应用程序添加安全控制，主要是指 OAuth 2.0。

（12）Spring Cloud Consul：封装了 Consul 操作。Consul 是一个服务发现与配置工具，与 Docker 可以无缝集成。

（13）Spring Cloud Zookeeper：操作 ZooKeeper 的工具包，用于使用 ZooKeeper 方式的服务注册和发现。

（14）Spring Cloud Stream：数据流操作开发包，封装了 Redis、RabbitMQ、Kafka 等发送、接收的消息。

（15）Spring Cloud CLI：基于 Spring Boot CLI，可以以命令行方式快速建立云组件。

Spring Cloud 从设计之初就考虑了绝大多数互联网公司架构演化所需的功能，因此 Spring Cloud 备受广大开发者的欢迎，本书也是基于 Spring Cloud 框架来完成微服务架构项目的。

1.2.2 Dubbo

Dubbo 是阿里巴巴公司开源的一个高性能、优秀的服务框架，使得应用可通过高性能的远程过程调用（Remote Procedure Call，RPC）实现服务的输出和输入功能，可以和 Spring 框架无缝集成。

Dubbo 是一款高性能、轻量级的开源 Java RPC 框架，其架构如图 1-8 所示。它提供的核心能力包括：面向接口的远程方法调用、智能容错和负载均衡、服务自动注册和发现。

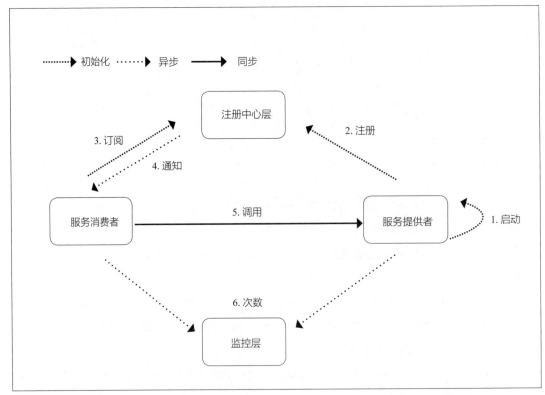

图1-8　Dubbo架构

服务提供者先启动服务，然后向注册中心注册服务。服务消费者订阅服务，如果订阅到自己想获得的服务，可以完成服务调用；如果没有订阅到自己想获得的服务，它会不断的尝试订阅。新的服务注册到注册中心以后，注册中心会将这些服务通知到服务消费者。

服务提供者和服务消费者通过异步的方式发送消息至监控平台，监控平台可以完成服务计数功能。

1.2.3 Dropwizard

Dropwizard 框架为开发者提供了一个非常简单的模型,里面有许多重要的模块,使用者可以根据需求添加业务逻辑,或者配置其他内容。该框架使用的 JAR 文件非常小,并且能够快速启动。

Dropwizard 最大的限制可能是缺乏依赖注入。如果希望使用依赖注入来保持代码的整洁和松耦合,则需要自己添加库,这点和 Spring 不同。现在 Dropwizard 也支持大多数功能,包括日志记录、健康检查和提供弹性代码等。

1.2.4 Cricket

Cricket 是一个用于快速开发 API 的框架。Cricket 很小,但它包括许多额外的功能,如键值数据存储,可避免连接数据库和调度程序控制后台重复处理。由于没有添加其他复杂的依赖项,因此你很容易将代码添加到 Cricket 并启动独立的微服务。

1.2.5 Jersey

开发 Web 服务的标准方法之一是 RESTful Web 服务的 Java API(又名 JAX-RS),这是 Jersey 框架中实现的通用方法。这种方法主要依赖于注释来指定路径映射和返回细节。从参数解析到 JSON 打包的所有其他内容都由 Jersey 处理。

Jersey 的主要优点是实现了 JAX-RS,这个特性非常受欢迎,因此一些开发人员习惯将 Jersey 与 Spring Boot 结合在一起使用。

1.2.6 Play

体验 JVM(Java 虚拟机)跨语言能力的最佳方式之一是使用 Play 框架,该框架是可以与 Java 或任何其他 JVM 语言兼容的。它的基础非常现代,具有异步、无状态的模型,不会让试图跟踪用户及其会话数据的线程使服务器过载。此外,Play 还有许多额外的特性可以用来充实网站,比如 OpenID、验证和文件上传支持。Play 代码库已经发展了 10 多年,因此使用者还会发现类似于对可扩展标记语言(Extensible Markup Language,XML)的支持这种"古老"的功能。Play 既成熟又轻盈,还比较有特色。

1.3 Spring Boot 和 Spring Cloud 的关系

在这里介绍 Spring Boot 是因为它是 Spring Cloud 的基础,其自身的各项优点,如自动化配置、快速开发、轻松部署等,使其非常适合作为微服务架构中各项具体微服务的开发框架。所以强烈推荐使用 Spring Boot 来构建微服务,它不仅可以帮助使用者快速地构建微服务,还可以轻松简单地整合 Spring Cloud 实现系统服务化。而如果使用传统的 Spring 构建方式,在整合过程中还需要做更多的依赖管理工作才能让微服务完好地运行起来。

1.3.1 认识Spring Boot

很多Spring框架的初学者经常会因为其繁杂的配置文件而却步。而很多老手每次构建新项目时总是会重复做复制一些差不多的配置文件等枯燥乏味的事。一名优秀的程序员或架构师，总会想尽办法来避免这样的重复劳动，比如，通过Maven等构建工具来创建针对不同场景的脚手架工程，在需要新建项目时通过这些脚手架来初始化自定义的标准工程，并根据需要做一些简单修改，以达到简化原有配置过程的效果。这样的做法虽然减少了工作量，但是这些配置依然大量散布在项目工程中，而大部分情况下使用者并不会去修改这些配置，那么这些配置为什么还要反复出现在项目工程中呢？实在有些碍眼！

Spring Boot的出现可以有效改善这类问题。Spring Boot的宗旨并非重写Spring或是替代Spring，而是通过设计大量的自动化配置等方式来简化Spring原有样板化的配置，使得开发者可以快速构建应用。Spring Boot官方简介如图1-9所示。

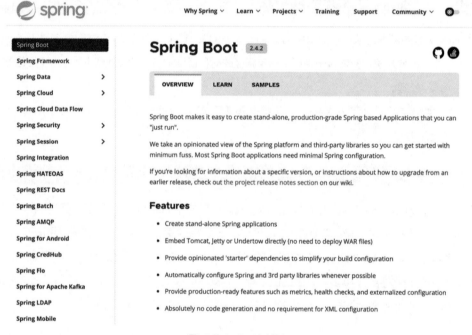

图1-9 Spring Boot官方简介

图1-9中"OVERVIEW"的部分描述翻译如下。

Spring Boot使创建独立的、基于Spring的生产级应用程序变得很容易，您可以"直接运行"。

我们认为，对于Spring平台和第三方库，您可以选择从最简单的开始。大多数Spring Boot应用程序需要最基础的Spring配置。

如果您正在寻找关于特定版本的信息，或关于如何从早期版本升级的说明，请查看我们在维基网上的项目版本说明部分。

除了解决配置问题，Spring Boot 还通过一系列 Starter 的定义，让使用者在整合各项功能的时候，不需要在 Maven 的 pom.xml 中维护那些错综复杂的依赖关系，而是通过类似模块化的 Starter 模块定义来引用，使得依赖管理工作变得更为简单。

在如今这个"容器化大行其道的时代"，Spring Boot 除了可以很好地融入 Docker，其自身还支持嵌入式的 Tomcat、Jetty 等容器。所以，通过 Spring Boot 构建的应用不再需要安装 Tomcat，将应用打成 WAR 包，再将其部署到 Tomcat 这样复杂的构建与部署动作，只需将 Spring Boot 应用打成 JAR 包，并通过 java -jar 命令直接运行就能启动一个标准化的 Web 应用，这使得 Spring Boot 应用变得非常轻便。

Spring Boot 对于构建、部署等做了这么多的优化，自然不能少了对开发环节的优化。

整个 Spring Boot 的生态系统都使用到了 Groovy。很自然地，使用者完全可以通过使用 Gradle 和 Groovy 来开发 Spring Boot 应用。比如下面的代码段，通过编译打包，执行 java -jar 命令就能启动一个返回"Hello World！"的 RESTful API。

```
@RestController
public class HWController {
  @RequestMapping(value = "/helloworld")
  public String helloWorld(){
     return "Hello World!";
  }
}
```

目前看来，Spring Cloud 是一套比较完整的微服务架构，它是一系列框架的有序集合。简单理解，它只是将目前开发得比较成熟、经得起实际考验的服务框架组合起来，通过 Spring Boot 进行再封装屏蔽掉了复杂的配置和实现原理，最终给开发者提供了一套简单易懂、易部署和易维护的分布式系统开发工具包。它利用 Spring Boot 的开发便利性巧妙地简化了分布式系统基础设施的开发，使服务发现注册、配置中心、消息总线、负载均衡、断路器、数据监控等，都可以用 Spring Boot 的开发风格做到一键启动和部署。

1.3.2 Spring Boot 整合 Spring Cloud

如果要将 Spring Cloud 添加到现有的 Spring Boot 应用程序，则第一步是确定要使用的 Spring Cloud 版本。在 Spring Cloud 的官网可以查看二者版本的对应关系，如图 1-10 所示。

Release Train	Boot Version
2020.0.x aka Ilford	2.4.x
Hoxton	2.2.x, 2.3.x (Starting with SR5)
Greenwich	2.1.x
Finchley	2.0.x
Edgware	1.5.x
Dalston	1.5.x

图1-10 Spring Cloud与Spring Boot版本的对应关系

Spring Cloud 不是一个组件，而是许多组件的集合。它的版本命名比较特殊，是由以 A 到 Z 为首字母的一些单词组成的。

Spring Boot 整合 Spring Cloud 需要在 POM 文件中添加对应版本的 Spring Cloud 的依赖，如下所示。

```xml
<parent>
    <groupId>org.springframework.boot</groupId>
    <artifactId>spring-boot-starter-parent</artifactId>
    <version>2.1.6.RELEASE</version>
</parent>
<dependencies>
<dependency>
    <groupId>org.springframework.cloud</groupId>
    <artifactId>spring-cloud-dependencies</artifactId>
    <version>Greenwich.RELEASE</version>
</dependency>
</dependencies>
```

这里 Spring Boot 使用的版本是 2.1.6，Spring Cloud 使用的版本是 Greenwich。

第 2 章

准备开发环境

工欲善其事,必先利其器。在学习本书的技术内容之前,应先将开发环境搭建好。本书所涉及的基础环境将在本章准备,包括 JDK、IntelliJ IDEA 和 Maven。如果读者对这些环境较为熟悉,可以直接跳过本章的内容。

本章的主要内容如下。

1. JDK 的下载、安装和 Java 环境变量的配置。
2. IntelliJ IDEA 的下载和安装。
3. Maven 的下载、安装、配置和与 IntelliJ IDEA 集成。

2.1 Java 开发环境 JDK

JDK 的英文全称是 Java Development Kit，即 Java 软件开发工具包，是 Oracle 公司提供的一套用于开发 Java 应用程序的开发包。它提供编译、运行 Java 应用程序所需要的各种工具和资源，包括 Java 编译器、Java 运行环境以及常用的 Java 类库等。

这里选择的是 Java SE 8，即 JDK 的一个标准版本，也是目前企业中使用较多的版本。

2.1.1 下载 JDK

访问 Oracle 的官网，滑动到底部，如图 2-1 所示。

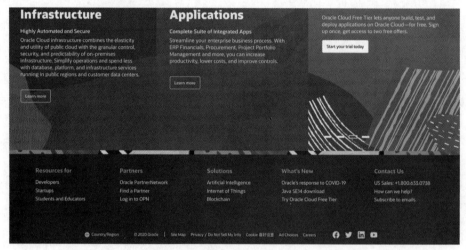

图2-1 Oracle官网

单击图 2-1 中的【Java SE14 download】跳转到下载页面，找到【Java SE 8】，如图 2-2 所示。

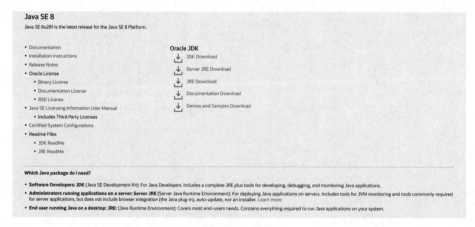

图2-2 Java SE 8

单击图 2-2 中的【JDK Download】跳转到下载选择页面，如图 2-3 所示。

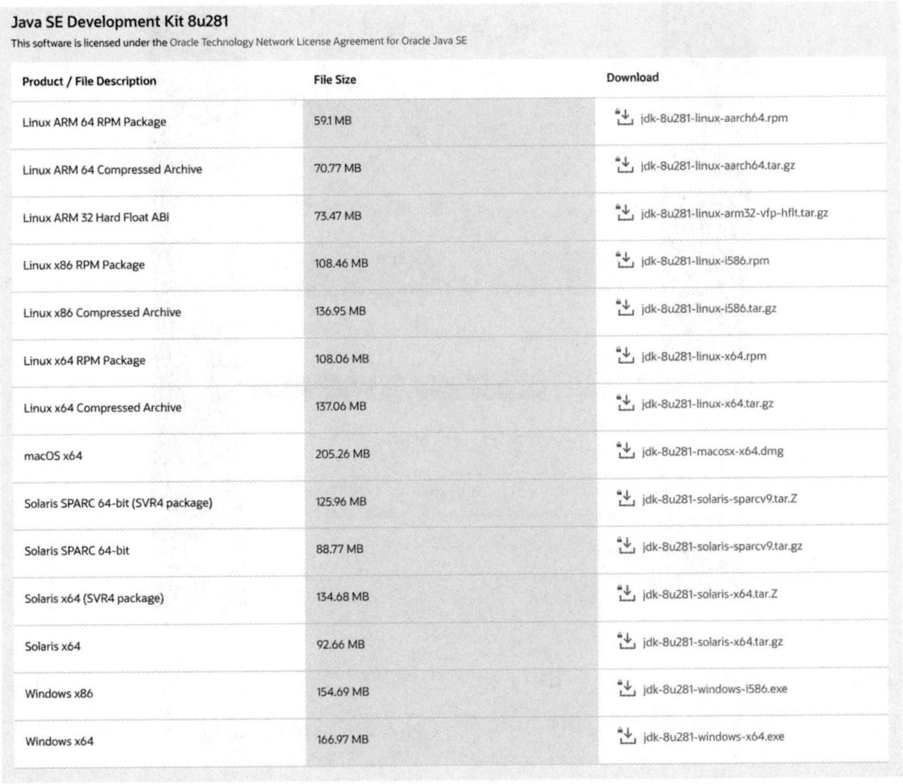

Java SE Development Kit 8u281

This software is licensed under the Oracle Technology Network License Agreement for Oracle Java SE

Product / File Description	File Size	Download
Linux ARM 64 RPM Package	59.1 MB	jdk-8u281-linux-aarch64.rpm
Linux ARM 64 Compressed Archive	70.77 MB	jdk-8u281-linux-aarch64.tar.gz
Linux ARM 32 Hard Float ABI	73.47 MB	jdk-8u281-linux-arm32-vfp-hflt.tar.gz
Linux x86 RPM Package	108.46 MB	jdk-8u281-linux-i586.rpm
Linux x86 Compressed Archive	136.95 MB	jdk-8u281-linux-i586.tar.gz
Linux x64 RPM Package	108.06 MB	jdk-8u281-linux-x64.rpm
Linux x64 Compressed Archive	137.06 MB	jdk-8u281-linux-x64.tar.gz
macOS x64	205.26 MB	jdk-8u281-macosx-x64.dmg
Solaris SPARC 64-bit (SVR4 package)	125.96 MB	jdk-8u281-solaris-sparcv9.tar.Z
Solaris SPARC 64-bit	88.77 MB	jdk-8u281-solaris-sparcv9.tar.gz
Solaris x64 (SVR4 package)	134.68 MB	jdk-8u281-solaris-x64.tar.Z
Solaris x64	92.66 MB	jdk-8u281-solaris-x64.tar.gz
Windows x86	154.69 MB	jdk-8u281-windows-i586.exe
Windows x64	166.97 MB	jdk-8u281-windows-x64.exe

图2-3　下载选择

在图 2-3 所示的页面中根据自己的操作系统选择对应的版本，本书中使用的是 Windows 7 64 位操作系统，单击【jdk-8u281-windows-x64.exe】进行下载，弹出对话框，勾选复选框，如图 2-4 所示。

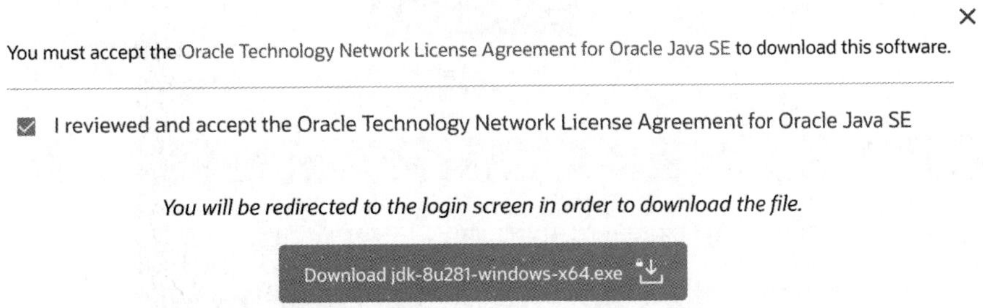

图2-4　下载

在图 2-4 所示的页面中单击【Download jdk-8u281-windows-x64.exe】跳转到 Oracle 的登录页面，如图 2-5 所示。

图2-5 登录

在图 2-5 所示的页面中输入用户名（图中"帐"应为"账"，后文同）和密码，单击【登录】按钮，然后等待下载完成即可。如果没有 Oracle 账户，可单击【创建账户】按钮，完成账户创建后再登录完成下载。下载完成后得到的软件的图标如图 2-6 所示。

图2-6 下载结果

2.1.2 安装 JDK

双击图 2-6 所示的软件图标进行安装，弹出安装向导提示窗口，如图 2-7 所示。

图2-7 向导提示

在图 2-7 所示的界面中单击【下一步】按钮,弹出选择安装位置对话框,如图 2-8 所示。

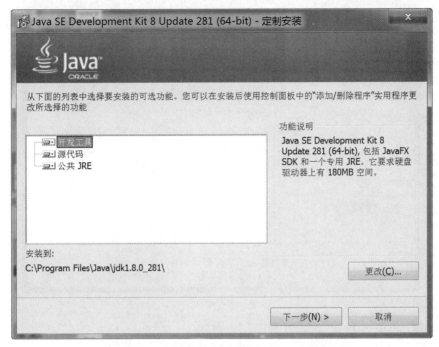

图2-8　选择安装位置

在图 2-8 所示的界面中单击【更改】按钮,选择安装的位置,本书选择安装到 C:\develop\Java\jdk1.8.0_281\。然后单击【下一步】按钮,弹出安装进度窗口,接着弹出 JRE 的安装窗口,如图 2-9 所示。

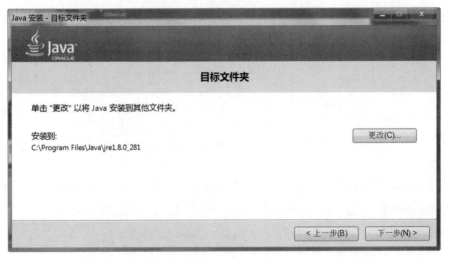

图2-9　安装JRE

因为安装的 JDK 中已经包含了 JRE,所以不需要安装 JRE。在图 2-9 所示的界面中单击右上角的【关闭】按钮,关闭此窗口,JDK 安装完成,如图 2-10 所示。

图2-10　安装完成

在图 2-10 所示的界面中单击【关闭】按钮，完成安装。

2.1.3　配置 Java 环境变量

右击【计算机】，单击【属性】→【高级系统设置】→【环境变量】，在系统变量下新建 JAVA_HOME 变量，其值是 JDK 的安装目录，如图 2-11 所示。

图2-11　新建系统变量JAVA_HOME

在图 2-11 所示的界面中单击【确定】按钮，然后找到系统变量 Path，单击【编辑】按钮，将光标移动到变量值文本框最前面，添加 %JAVA_HOME%\bin;，记得使用分号，如图 2-12 所示。

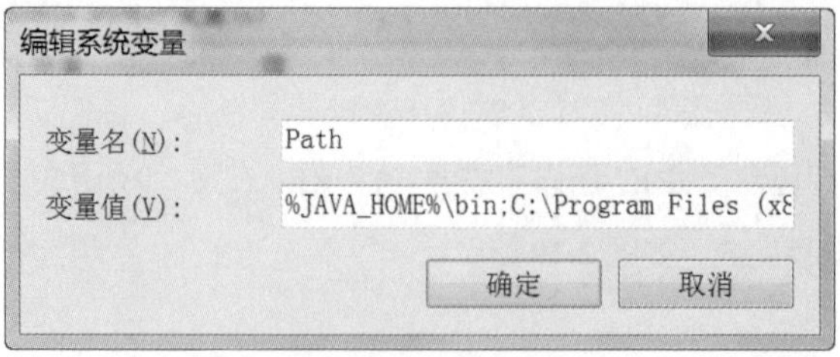

图2-12　添加Java环境变量

在图 2-12 所示的界面中单击【确定】按钮，完成设置。

打开命令行界面，执行 java -version 命令，显示 Java 的版本信息，如图 2-13 所示。

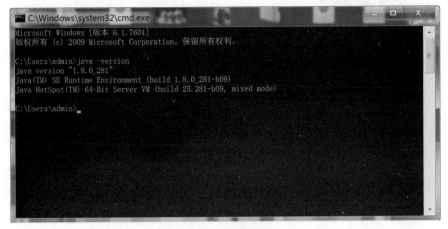

图2-13　Java的版本信息

至此，JDK 安装完成。

 开发工具 IntelliJ IDEA

IntelliJ IDEA 的各个方面都旨在最大程度地提高开发人员的生产力，其智能编码辅助和人体工程学设计不仅使开发富有成效，而且过程令人愉悦。

2.2.1　下载 IntelliJ IDEA

访问 IntelliJ IDEA 的官网，如图 2-14 所示。

图2-14　IntelliJ IDEA官网

在图 2-14 所示的页面中单击【Download】按钮，跳转到下载页面，如图 2-15 所示。

图2-15 下载

在图 2-15 所示的页面中根据自己的操作系统选择对应的版本，本书中使用的是 Windows 7 64 位操作系统，单击【Windows】，单击【Download】按钮进行下载。下载完成后得到的软件的图标如图 2-16 所示。

图2-16 下载结果

2.2.2 安装 IntelliJ IDEA

在图 2-16 所示的界面中双击该软件图标进行安装，弹出欢迎信息窗口，如图 2-17 所示。

图2-17 欢迎信息

在图 2-17 所示的界面中单击【 Next 】按钮，弹出选择安装位置窗口，单击【 Browse 】按钮选择安装的位置，本书中选择安装到 C:\develop\IntelliJIDEA2020.3.1，如图 2-18 所示。

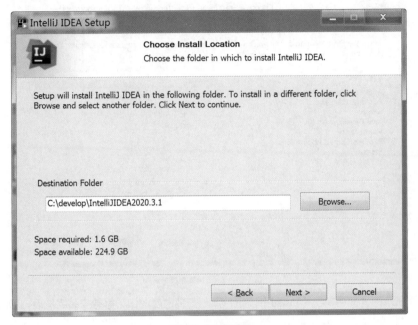

图2-18　选择安装位置

在图 2-18 所示的界面中单击【 Next 】按钮，弹出安装选项窗口，如图 2-19 所示。

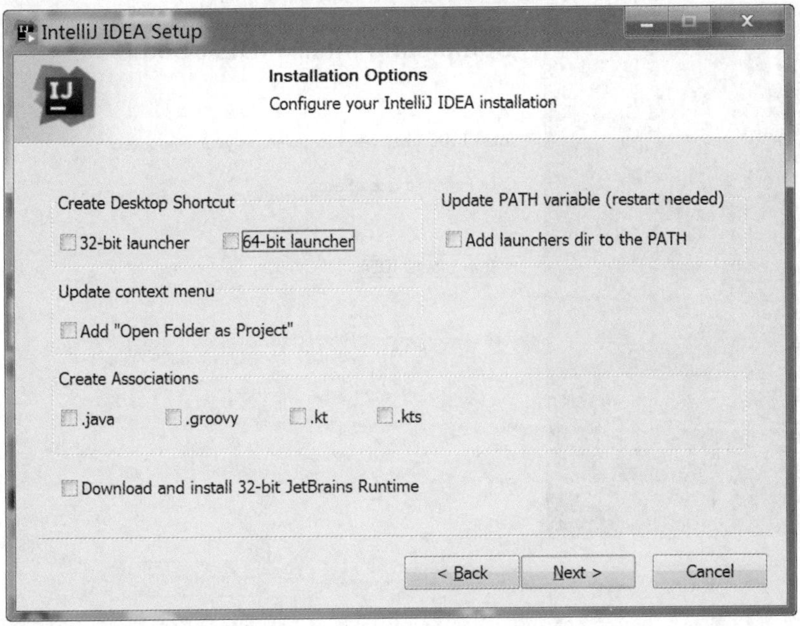

图2-19　安装选项

在图 2-19 所示的界面中单击【 Next 】按钮，弹出选择开始菜单文件夹窗口，如图 2-20 所示。

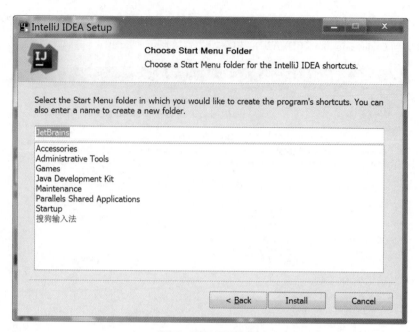

图2-20　选择开始菜单文件夹

在图 2-20 所示的界面中单击【Install】按钮，弹出安装进度窗口，等待安装完成后，弹出提示安装完成窗口，如图 2-21 所示。

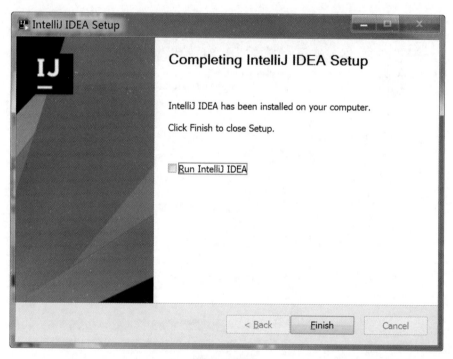

图2-21　安装完成

在图 2-21 所示的界面中单击【Finish】按钮，完成安装。

通过开始菜单找到 IntelliJ IDEA，单击启动后，弹出用户同意对话框，勾选复选框，如图 2-22 所示。

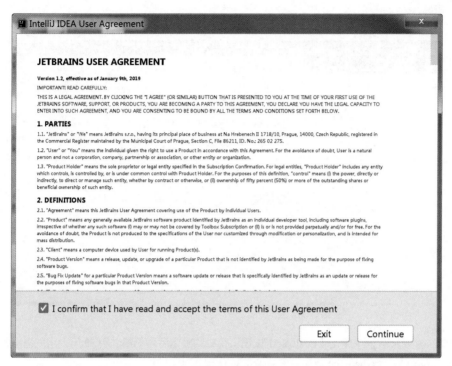

图2-22　用户同意

在图 2-22 所示的界面中单击【Continue】按钮，弹出数据分享对话框，如图 2-23 所示。

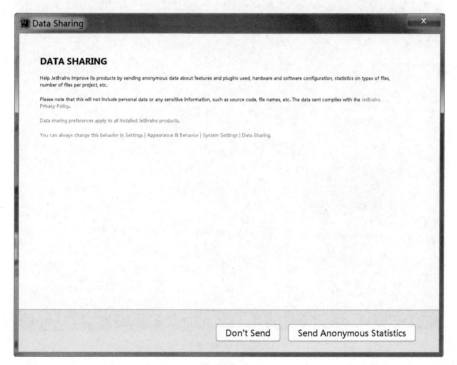

图2-23　数据分享

在图 2-23 所示的界面中，单击【Don't Send】按钮，弹出激活窗口，选中【Evaluate for free】，如图 2-24

所示。

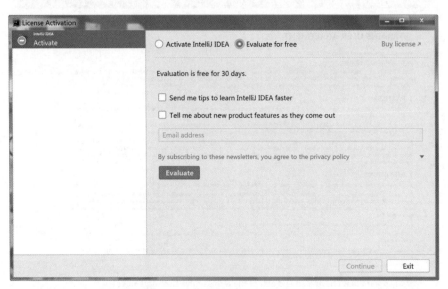

图2-24 激活

在图 2-24 所示的界面中单击【Evaluate】按钮（完成此操作后可免费试用 30 天，用户也可以根据自己的需要购买激活码），接着弹出欢迎界面，如图 2-25 所示。

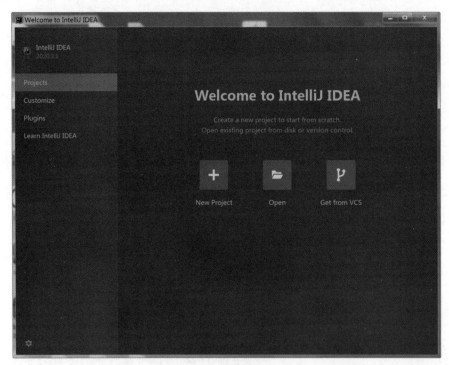

图2-25 欢迎界面

在图 2-25 所示的界面中单击【New Project】创建项目，弹出新项目对话框，用户可以根据自己的需求选择项目的类型和选项，如图 2-26 所示。

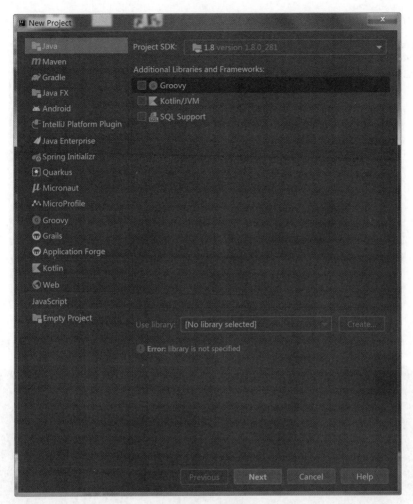

图2-26　创建项目

在图 2-26 所示的界面中单击【Next】按钮,弹出创建模板项目对话框,如图 2-27 所示。

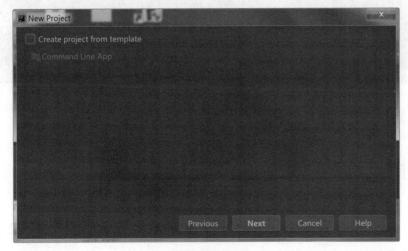

图2-27　创建模板项目

在图 2-27 所示的界面中单击【Next】按钮，弹出新项目对话框，在这里可以设置项目的名称和安装路径，如图 2-28 所示。

图2-28　设置项目名称和安装路径

在图 2-28 所示的界面中单击【Finish】按钮，弹出项目界面，创建项目完成，如图 2-29 所示。

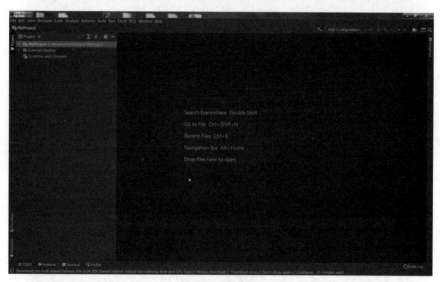

图2-29　创建项目完成

至此，开发工具 IntelliJ IDEA 安装完成。

2.3　项目管理工具 Maven

Maven 是一个项目管理工具，它包含了一个项目对象模型、一组标准集合、一个项目生命周期、一个依赖管理系统和用来运行定义在生命周期阶段中插件目标的逻辑。

2.3.1 下载 Maven

访问 Maven 的官网，如图 2-30 所示。

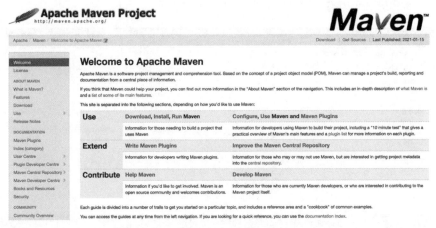

图2-30　Maven官网

在图 2-30 所示的界面左侧单击【Download】，跳转到下载页面，如图 2-31 所示。

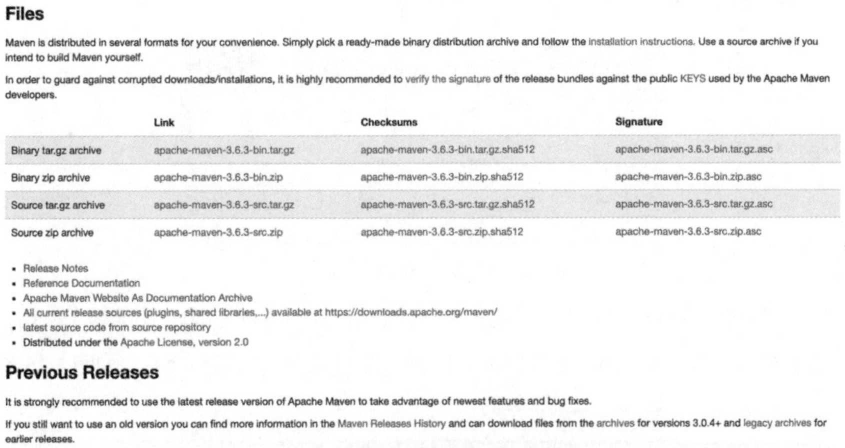

图2-31　下载

在图 2-31 所示的界面中根据自己的操作系统选择对应的版本，本书中使用的是 Windows 7 64 位操作系统，单击【apache-maven-3.6.3-bin.zip】进行下载，下载完成后得到的软件压缩包如图 2-32 所示。

图2-32　下载结果

2.3.2 安装 Maven

将图 2-32 中的压缩包解压到指定位置，本书选择解压到 C:\develop\apache-maven-3.6.3（注意去掉多余的内层目录），此目录下可以直接找到 bin 文件夹，如图 2-33 所示。

图2-33 安装Maven

2.3.3 配置 Maven 环境变量

右击【计算机】，单击【属性】→【高级系统设置】→【环境变量】，在系统变量下新建 MAVEN_HOME 变量，其值是 MAVEN 的安装目录，如图 2-34 所示。

图2-34 新建系统变量MAVEN_HOME

在图 2-34 所示的界面中单击【确定】按钮，然后找到系统变量 Path，单击【编辑】按钮，将光标移动到变量值文本框最前面，添加 %MAVEN_HOME%\bin，记得使用分号与后面隔开，如图 2-35 所示。

图2-35 添加Maven环境变量

在图 2-35 所示的界面中单击【确定】按钮，完成设置。

打开命令行界面，执行 mvn -v 命令，显示 Maven 的版本信息，如图 2-36 所示。

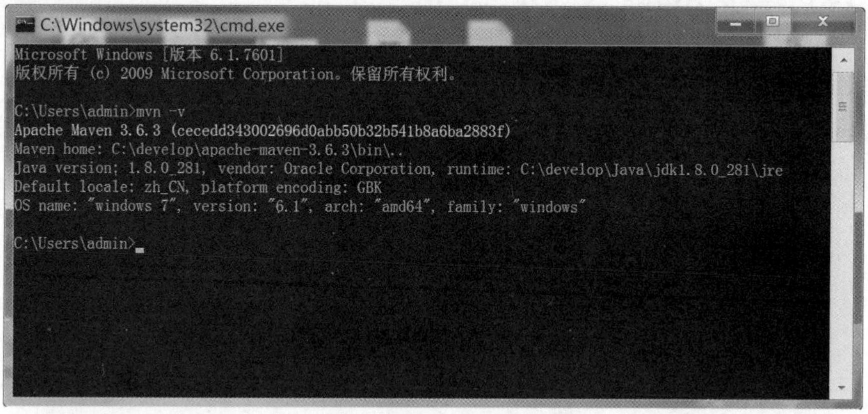

图2-36 Maven的版本信息

至此，Maven 安装完成。

2.3.4 配置 Maven 本地仓库和下载源

在 %MAVEN_HOME%/conf/settings.xml 文件中配置本地仓库位置。打开 settings.xml 文件，配置信息如图 2-37 所示。

```
<settings xmlns="http://maven.apache.org/SETTINGS/1.0.0"
          xmlns:xsi="http://www.w3.org/2001/XMLSchema-instance"
          xsi:schemaLocation="http://maven.apache.org/SETTINGS/1.0.0 http://maven.apache.org/xsd/settings-1.0.0.xsd">
    <!-- localRepository
     | The path to the local repository maven will use to store artifacts.
     |
     | Default: ${user.home}/.m2/repository
    <localRepository>/path/to/local/repo</localRepository>
    -->
    <localRepository>C:/develop/maven_repository</localRepository>
    <!-- interactiveMode
     | This will determine whether maven prompts you when it needs input. If set to false,
     | maven will use a sensible default value, perhaps based on some other setting, for
     | the parameter in question.
     |
     | Default: true
    <interactiveMode>true</interactiveMode>
    -->
```

图2-37 配置Maven本地仓库位置

同时在本地创建目录 C:/develop/maven_repository。

在 %MAVEN_HOME%/conf/settings.xml 文件中配置 Maven 下载源，本书中配置 Maven 使用阿里云的下载源。打开 settings.xml 文件，配置信息如图 2-38 所示。

```
<mirrors>
    <!-- mirror
     | Specifies a repository mirror site to use instead of a given repository. The repository that
     | this mirror serves has an ID that matches the mirrorOf element of this mirror. IDs are used
     | for inheritance and direct lookup purposes, and must be unique across the set of mirrors.
     |
    <mirror>
      <id>mirrorId</id>
      <mirrorOf>repositoryId</mirrorOf>
      <name>Human Readable Name for this Mirror.</name>
      <url>http://my.repository.com/repo/path</url>
    </mirror>
     -->
    <mirror>
        <id>alimaven</id>
        <name>aliyun maven</name>
        <url>http://maven.aliyun.com/nexus/content/groups/public/</url>
        <mirrorOf>central</mirrorOf>
    </mirror>
</mirrors>
```

图2-38 配置Maven使用阿里云下载源

2.3.5 与 IntelliJ IDEA 集成

打开 IntelliJ IDEA，单击【File】→【settings】→【Build,Execution,Deployment】→【Build Tools】→【Maven】，找到 Maven 的配置选项，如图 2-39 所示。

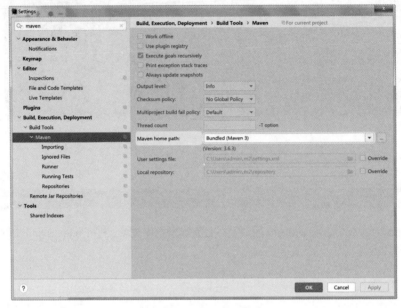

图2-39 Maven配置选项

在图 2-39 所示的界面中，单击【Maven home path】右侧的选项选择 Maven 的安装路径；单击【User settings file】右侧的【Override】，选择 Maven 的配置文件 settings.xml；单击【Local repository】右侧的【Override】，选择 Maven 的仓库路径，结果如图 2-40 所示。然后单击【OK】按钮，完成 Maven 与 IntelliJ IDEA 的集成。

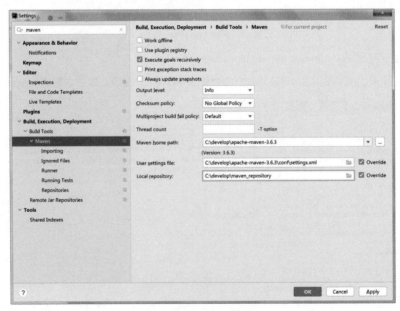

图2-40 配置Maven本地信息

第3章 贯穿案例

使用微服务架构的分布式系统时,微服务之间通过网络通信,通过服务提供者与服务消费者来描述微服务间的调用关系。

本章以电商系统中常见的用户下单为例。用户向订单微服务发起一个购买的请求,在保存订单之前需要调用商品微服务查询当前商品库存、单价等信息。在这种场景下,订单微服务就是一个服务消费者,商品微服务就是一个服务提供者。

本章的主要内容如下。

1. 数据库。
2. 创建工程。
3. 使用 Postman 测试微服务。
4. 调用微服务。

3.1 数据库

这里使用的数据库是 MySQL 5.7，数据库包含 3 张表，分别是用户表、商品表和订单表。在分布式项目中，一般业务比较复杂，数据量也比较大，每个微服务都有自己的数据库，每个数据库中都有多张表。为了方便介绍，本书中的贯穿案例有 3 个微服务，3 个微服务使用的都是同一个数据库，每个微服务分别对应一张表。数据库层使用的是 Spring Data JPA 框架。

用户表的代码如下所示。

```sql
CREATE TABLE 'tb_user'(
'id' INT (11) NOT NULL AUTO_INCREMENT,
'name' VARCHAR (40) DEFAULT NULL COMMENT '用户名',
'password' VARCHAR (40) DEFAULT NULL COMMENT '密码',
'age' INT (3) DEFAULT NULL COMMENT '年龄',
'balance' DECIMAL (10,2) DEFAULT NULL COMMENT '余额',
'address' VARCHAR (80) DEFAULT NULL COMMENT '地址',
PRIMARY KEY ('id')
) ENGINE = InnoDB DEFAULT CHARSET = utf8;
```

商品表的代码如下所示。

```sql
CREATE TABLE 'tb_product' (
'id' INT (11) NOT NULL AUTO_INCREMENT,
'product_name' VARCHAR (40) DEFAULT NULL COMMENT '名称',
'status' INT (2) DEFAULT NULL COMMENT '状态',
'price' DECIMAL (10,2) DEFAULT NULL COMMENT '单价',
'product_desc' VARCHAR (255) DEFAULT NULL COMMENT '描述',
'caption' VARCHAR (255) DEFAULT NULL COMMENT '标题',
'inventory' INT (11) DEFAULT NULL COMMENT '库存',
PRIMARY KEY ('id')
) ENGINE = InnoDB DEFAULT CHARSET = utf8;
```

订单表的代码如下所示。

```sql
CREATE TABLE 'tb_order' (
'id' INT (11) NOT NULL AUTO_INCREMENT,
'user_id' INT (11) DEFAULT NULL COMMENT '用户id',
'product_id' INT (11) DEFAULT NULL COMMENT '商品id',
'number' INT (11) DEFAULT NULL COMMENT '数量',
'price' DECIMAL (10,2) DEFAULT NULL COMMENT '单价',
'amount' DECIMAL (10,2) DEFAULT NULL COMMENT '总额',
'product_name' VARCHAR (40) DEFAULT NULL COMMENT '商品名',
'user_name' VARCHAR (40) DEFAULT NULL COMMENT '用户名',
PRIMARY KEY ('id')
) ENGINE = InnoDB DEFAULT CHARSET = utf8;
```

3.2 创建工程

整个工程的结构如图 3-1 所示。先创建父工程 shop_parent，然后创建子工程（用户微服务 user_service、商品微服务 goods_service 和订单微服务 order_service）。

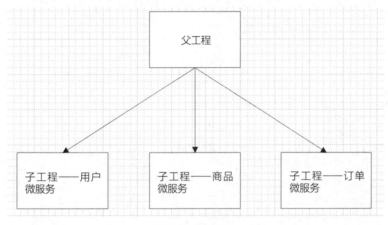

图3-1 整个工程的结构

3.2.1 创建父工程

使用 IntelliJ IDEA 创建父工程 shop_parent，如图 3-2 所示。

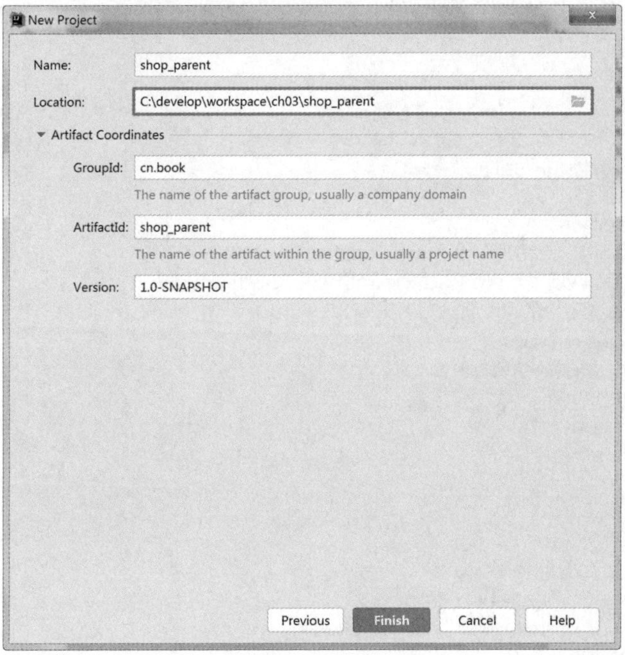

图3-2 创建父工程

创建父工程后,在其对应的 pom.xml 文件中导入 Maven 坐标,如下所示。

```xml
<parent>
    <groupId>org.springframework.boot</groupId>
    <artifactId>spring-boot-starter-parent</artifactId>
    <version>2.1.6.RELEASE</version>
</parent>

<properties>
    <project.build.sourceEncoding>UTF-8</project.build.sourceEncoding>
    <project.reporting.outputEncoding>UTF-8</project.reporting.outputEncoding>
    <java.version>1.8</java.version>
</properties>

<dependencies>
    <dependency>
        <groupId>org.springframework.boot</groupId>
        <artifactId>spring-boot-starter-web</artifactId>
    </dependency>
    <dependency>
        <groupId>org.springframework.boot</groupId>
        <artifactId>spring-boot-starter-logging</artifactId>
    </dependency>
    <dependency>
        <groupId>org.springframework.boot</groupId>
        <artifactId>spring-boot-starter-test</artifactId>
        <scope>test</scope>
    </dependency>
    <dependency>
        <groupId>org.projectlombok</groupId>
        <artifactId>lombok</artifactId>
        <version>1.18.4</version>
        <scope>provided</scope>
    </dependency>
</dependencies>


<dependencyManagement>
    <dependencies>
        <dependency>
            <groupId>org.springframework.cloud</groupId>
            <artifactId>spring-cloud-dependencies</artifactId>
            <version>Greenwich.RELEASE</version>
            <type>pom</type>
            <scope>import</scope>
        </dependency>
    </dependencies>
</dependencyManagement>

<repositories>
    <repository>
        <id>spring-snapshots</id>
        <name>Spring Snapshots</name>
        <url>http://repo.spring.io/libs-snapshot-local</url>
        <snapshots>
```

```xml
        <enabled>true</enabled>
      </snapshots>
    </repository>
    <repository>
      <id>spring-milestones</id>
      <name>Spring Milestones</name>
      <url>http://repo.spring.io/libs-milestone-local</url>
      <snapshots>
        <enabled>false</enabled>
      </snapshots>
    </repository>
    <repository>
      <id>spring-releases</id>
      <name>Spring Releases</name>
      <url>http://repo.spring.io/libs-release-local</url>
      <snapshots>
        <enabled>false</enabled>
      </snapshots>
    </repository>
  </repositories>
  <pluginRepositories>
    <pluginRepository>
      <id>spring-snapshots</id>
      <name>Spring Snapshots</name>
      <url>http://repo.spring.io/libs-snapshot-local</url>
      <snapshots>
        <enabled>true</enabled>
      </snapshots>
    </pluginRepository>
    <pluginRepository>
      <id>spring-milestones</id>
      <name>Spring Milestones</name>
      <url>http://repo.spring.io/libs-milestone-local</url>
      <snapshots>
        <enabled>false</enabled>
      </snapshots>
    </pluginRepository>
  </pluginRepositories>

  <build>
    <plugins>
      <plugin>
        <groupId>org.springframework.boot</groupId>
        <artifactId>spring-boot-maven-plugin</artifactId>
      </plugin>
    </plugins>
  </build>
```

3.2.2 创建子工程——用户微服务

创建子工程

使用 IntelliJ IDEA 创建子工程——用户微服务，如图 3-3 所示。

图3-3 用户微服务

创建子工程后,在其对应的 pom.xml 文件中导入 Maven 坐标,如下所示。

```
<dependency>
    <groupId>mysql</groupId>
    <artifactId>mysql-connector-java</artifactId>
    <version>5.1.32</version>
</dependency>
<dependency>
    <groupId>org.springframework.boot</groupId>
    <artifactId>spring-boot-starter-data-jpa</artifactId>
</dependency>
```

创建用户实体类

创建用户微服务相对应的用户实体类,加入 JPA 注解和 Lombok 注解,如下所示。

```
package cn.book.user.entity;

import lombok.Data;

import javax.persistence.*;
import java.math.BigDecimal;

/**
 * 用户实体类
 */
@Data
@Entity
```

```java
@Table(name="tb_user")
public class User {

  @Id
@GeneratedValue(strategy = GenerationType.IDENTITY)
    private Long id;
    private String name;
    private String password;
    private Integer age;
    private BigDecimal balance;
    private String address;
}
```

■ 创建 Dao 层接口

创建用户微服务相对应的 Dao 层接口，继承 JPA 的接口，自动实现基本的新增、删除、修改、查询（以下简称增删改查）功能，如下所示。

```java
package cn.book.user.dao;

import cn.book.user.entity.User;
import org.springframework.data.jpa.repository.JpaRepository;
import org.springframework.data.jpa.repository.JpaSpecificationExecutor;

public interface UseDao extends JpaRepository<User,Long>, JpaSpecificationExecutor<User> {

}
```

■ 创建 Service 层接口

创建用户微服务相对应的 Service 层接口，如下所示。

```java
package cn.book.user.service;

import cn.book.user.entity.User;

import java.util.List;

public interface UserService {
    // 根据 id 查询单个
    User findById(Long id);

    // 查询全部
    List findAll();

    // 新增
    void save(User user);

    // 修改
    void update(User user);

    // 删除
    void delete(Long id);
}
```

创建用户微服务相对应的 Service 层接口实现,并注入 Dao 层对象,如下所示。

```
package cn.book.user.service.impl;

import cn.book.user.dao.UserDao;
import cn.book.user.entity.User;
import cn.book.user.service.UserService;
import org.springframework.beans.factory.annotation.Autowired;

import java.util.List;

public class UserServiceImpl implements UserService {
    @Autowired
    private UserDao userDao;

    @Override
    public User findById(Long id) {
        return userDao.findById(id).get();
    }

    @Override
    public List findAll() {
        return userDao.findAll();
    }

    @Override
    public void save(User user) {
        userDao.save(user);
    }

    @Override
    public void update(User user) {
        userDao.save(user);
    }

    @Override
    public void delete(Long id) {
        userDao.deleteById(id);
    }
}
```

■ 创建 Web 层控制器

创建用户微服务相对应的 Web 层控制器,并注入 Service 层对象,如下所示。

```
package cn.book.user.controller;

import cn.book.user.entity.User;
import cn.book.user.service.UserService;
import org.springframework.beans.factory.annotation.Autowired;
import org.springframework.beans.factory.annotation.Value;
import org.springframework.web.bind.annotation.*;

import java.util.List;
```

```java
@RestController
@RequestMapping("/user")
public class UserController {

    @Autowired
    private UserService userService;

    @RequestMapping(value = "/{id}", method = RequestMethod.GET)
    public User findById(@PathVariable Long id) {
        return userService.findById(id);
    }

    @RequestMapping(value = "/findAll", method = RequestMethod.GET)
    public List findAll() {
        return userService.findAll();
    }

    @RequestMapping(value = "/save", method = RequestMethod.POST)
    public String save(@RequestBody User user) {
        userService.save(user);
        return " 保存成功 ";
    }

    @RequestMapping(value = "/{id}", method = RequestMethod.PUT)
    public String update(@RequestBody User user, @PathVariable Long id) {
        user.setId(id);
        userService.update(user);
        return " 修改成功 ";
    }

    @RequestMapping(value = "/{id}", method = RequestMethod.DELETE)
    public String deleteById(@PathVariable Long id) {
        userService.delete(id);
        return " 删除成功 ";
    }
}
```

创建启动类

创建用户微服务相对应的启动类 User Application，如下所示。

```java
package cn.book.user;

import org.springframework.boot.SpringApplication;
import org.springframework.boot.autoconfigure.SpringBootApplication;
import org.springframework.boot.autoconfigure.domain.EntityScan;

@SpringBootApplication
@EntityScan("cn.book.user.entity")
public class UserApplication {

    public static void main(String[] args) {
        SpringApplication.run(UserApplication.class, args);
    }
}
```

■ 创建配置文件

创建用户微服务相对应的 YML 配置文件，如下所示。

```yml
server:
  port: 9001 # 端口
spring:
  application:
    name: user_service # 服务名称
  datasource: # 数据源
    driver-class-name: com.mysql.jdbc.Driver
    url: jdbc:mysql://192.168.10.167:3306/shop?characterEncoding=utf8
    username: root
    password: root
  jpa:
    database: MySQL
    show-sql: true
    open-in-view: true
```

至此，用户微服务创建成功，可以运行 User Application 进行访问。

3.2.3 创建子工程——商品微服务

■ 创建子工程

使用 IntelliJ IDEA 创建子工程——商品微服务，如图 3-4 所示。

图3-4 商品微服务

创建子工程后，在其对应的 pom.xml 文件中导入 Maven 坐标，如下所示。

```xml
<dependency>
    <groupId>mysql</groupId>
    <artifactId>mysql-connector-java</artifactId>
    <version>5.1.32</version>
</dependency>
<dependency>
    <groupId>org.springframework.boot</groupId>
    <artifactId>spring-boot-starter-data-jpa</artifactId>
</dependency>
```

创建商品实体类

创建商品微服务相对应的商品实体类，加入 JPA 注解和 Lombok 注解，如下所示。

```java
package cn.book.product.entity;

import lombok.Data;

import javax.persistence.*;
import java.math.BigDecimal;

/**
 * 商品实体类
 */
@Data
@Entity
@Table(name = "tb_product")
public class Product {

    @Id
    @GeneratedValue(strategy = GenerationType.IDENTITY)
    private Long id;
    private String product_name;
    private Integer status;
    private BigDecimal price;
    private String product_desc;
    private String caption;
    private Integer inventory;
}
```

创建 Dao 层接口

创建商品微服务相对应的 Dao 层接口，继承 JPA 的接口，自动实现基本的增删改查功能，如下所示。

```java
package cn.book.product.dao;

import cn.book.product.entity.Product;
import org.springframework.data.jpa.repository.JpaRepository;
import org.springframework.data.jpa.repository.JpaSpecificationExecutor;

public interface ProductDao extends JpaRepository<Product, Long>, JpaSpecificationExecutor<Product> {
```

}

创建 Service 层接口

创建商品微服务相对应的 Service 层接口，如下所示。

```java
package cn.book.product.service;

import cn.book.product.entity.Product;

import java.util.List;

public interface ProductService {
    // 根据 id 查询单个
    Product findById(Long id);

    // 查询全部
    List findAll();

    // 新增
    void save(Product product);

    // 修改
    void update(Product product);

    // 删除
    void delete(Long id);
}
```

创建商品微服务相对应的 Service 层接口实现，并注入 Dao 层对象，如下所示。

```java
package cn.book.product.service.impl;

import cn.book.product.dao.ProductDao;
import cn.book.product.entity.Product;
import cn.book.product.service.ProductService;
import org.springframework.beans.factory.annotation.Autowired;
import org.springframework.stereotype.Service;

import java.util.List;

@Service
public class ProductServiceImpl implements ProductService {
    @Autowired
    private ProductDao productDao;

    @Override
    public Product findById(Long id) {
        return productDao.findById(id).get();
    }

    @Override
    public List findAll() {
        return productDao.findAll();
    }
```

```java
    @Override
    public void save(Product product) {
        productDao.save(product);
    }

    @Override
    public void update(Product product) {
        productDao.save(product);
    }

    @Override
    public void delete(Long id) {
        productDao.deleteById(id);
    }
}
```

创建 Web 层控制器

创建商品微服务相对应的 Web 层控制器，并注入 Service 层对象，如下所示。

```java
package cn.book.product.controller;

import cn.book.product.entity.Product;
import cn.book.product.service.ProductService;
import org.springframework.beans.factory.annotation.Autowired;
import org.springframework.web.bind.annotation.*;

import java.util.List;

@RestController
@RequestMapping("/product")
public class ProductController {

    @Autowired
    private ProductService productService;

    @RequestMapping(value = "/{id}", method = RequestMethod.GET)
    public Product findById(@PathVariable Long id) {
        return productService.findById(id);
    }

    @RequestMapping(value = "/findAll", method = RequestMethod.GET)
    public List findAll() {
        return productService.findAll();
    }

    @RequestMapping(value = "/save", method = RequestMethod.POST)
    public String save(@RequestBody Product product) {
        productService.save(product);
        return "保存成功";
    }

    @RequestMapping(value = "/{id}", method = RequestMethod.PUT)
    public String update(@RequestBody Product product,@PathVariable Long id) {
```

```
        product.setId(id);
        productService.update(product);
        return " 修改成功 ";
    }

    @RequestMapping(value = "/{id}", method = RequestMethod.DELETE)
    public String deleteById(@PathVariable Long id) {
        productService.delete(id);
        return " 删除成功 ";
    }
}
```

■ **创建启动类**

创建商品微服务相对应的启动类 Product Application，如下所示。

```
package cn.book.product;

import org.springframework.boot.SpringApplication;
import org.springframework.boot.autoconfigure.SpringBootApplication;
import org.springframework.boot.autoconfigure.domain.EntityScan;


@SpringBootApplication
@EntityScan("cn.book.product.entity")
public class ProductApplication {

    public static void main(String[] args) {
        SpringApplication.run(ProductApplication.class, args);
    }
}
```

■ **创建配置文件**

创建商品微服务相对应的 YML 配置文件，如下所示。

```
server:
  port: 9002 # 端口
spring:
  application:
    name: product_service # 服务名称
  datasource: # 数据源
    driver-class-name: com.mysql.jdbc.Driver
    url: jdbc:mysql://192.168.10.167:3306/shop?characterEncoding=utf8
    username: root
    password: root
  jpa:
    database: MySQL
    show-sql: true
    open-in-view: true
```

至此，商品微服务创建成功，可以运行 Product Application 进行访问。

3.2.4 创建子工程——订单微服务

创建子工程

使用 IntelliJ IDEA 创建子工程——订单微服务,如图 3-5 所示。

图3-5 订单微服务

创建子工程后,在其对应的 pom.xml 文件中导入 Maven 坐标,如下所示。

```
<dependency>
    <groupId>mysql</groupId>
    <artifactId>mysql-connector-java</artifactId>
    <version>5.1.32</version>
</dependency>
<dependency>
    <groupId>org.springframework.boot</groupId>
    <artifactId>spring-boot-starter-data-jpa</artifactId>
</dependency>
```

创建订单实体类

创建订单微服务相对应的订单实体类,加入 JPA 注解和 Lombok 注解,如下所示。

```
package cn.book.order.entity;
```

```
import lombok.Data;

import javax.persistence.*;
import java.math.BigDecimal;

/**
 * 订单实体类
 */
@Data
@Entity
@Table(name = "tb_order")
public class Order {

    @Id
    @GeneratedValue(strategy = GenerationType.IDENTITY)
    private Long id;
    private Long userId;
    private Long productId;
    private Integer number;
    private BigDecimal price;
    private BigDecimal amount;
    private String productName;
    private String userName;
}
```

■ 创建 Dao 层接口

创建订单微服务相对应的 Dao 层接口，继承 JPA 的接口，自动实现基本的增删改查，如下所示。

```
package cn.book.order.dao;

import cn.book.order.entity.Order;
import org.springframework.data.jpa.repository.JpaRepository;
import org.springframework.data.jpa.repository.JpaSpecificationExecutor;

public interface OrderDao extends JpaRepository<Order, Long>, JpaSpecificationExecutor<Order>
{

}
```

■ 创建 Service 层接口

创建订单微服务相对应的 Service 层接口，如下所示。

```
package cn.book.order.service;

import cn.book.order.entity.Order;

import java.util.List;

public interface OrderService {
    // 根据 id 查询单个
    Order findById(Long id);
```

```java
    // 查询全部
    List findAll();

    // 新增
    void save(Order order);

    // 修改
    void update(Order order);

    // 删除
    void delete(Long id);
}
```

创建订单微服务相对应的 Service 层接口实现，并注入 Dao 层对象，如下所示。

```java
package cn.book.order.service.impl;

import cn.book.order.dao.OrderDao;
import cn.book.order.entity.Order;
import cn.book.order.service.OrderService;
import org.springframework.beans.factory.annotation.Autowired;
import org.springframework.stereotype.Service;

import java.util.List;

@Service
public class OrderServiceImpl implements OrderService {
    @Autowired
    private OrderDao orderDao;

    @Override
    public Order findById(Long id) {
        return orderDao.findById(id).get();
    }

    @Override
    public List findAll() {
        return orderDao.findAll();
    }

    @Override
    public void save(Order order) {
        orderDao.save(order);
    }

    @Override
    public void update(Order order) {
        orderDao.save(order);
    }

    @Override
    public void delete(Long id) {
        orderDao.deleteById(id);
    }
}
```

■ 创建 Web 层控制器

创建订单微服务相对应的 Web 层控制器,并注入 Service 层对象,如下所示。

```java
package cn.book.order.controller;

import cn.book.order.entity.Order;
import cn.book.order.service.OrderService;
import org.springframework.beans.factory.annotation.Autowired;
import org.springframework.web.bind.annotation.*;

import java.util.List;

@RestController
@RequestMapping("/order")
public class OrderController {

    @Autowired
    private OrderService orderService;

    @RequestMapping(value = "/{id}", method = RequestMethod.GET)
    public Order findById(@PathVariable Long id) {
        return orderService.findById(id);
    }

    @RequestMapping(value = "/findAll", method = RequestMethod.GET)
    public List findAll() {
        return orderService.findAll();
    }

    @RequestMapping(value = "/save", method = RequestMethod.POST)
    public String save(@RequestBody Order order) {
        orderService.save(order);
        return " 保存成功 ";
    }

    @RequestMapping(value = "/{id}", method = RequestMethod.PUT)
    public String update(@RequestBody Order order,@PathVariable Long id) {
        order.setId(id);
        orderService.update(order);
        return " 修改成功 ";
    }

    @RequestMapping(value = "/{id}", method = RequestMethod.DELETE)
    public String deleteById(@PathVariable Long id) {
        orderService.delete(id);
        return " 删除成功 ";
    }
}
```

■ 创建启动类

创建订单微服务相对应的启动类 Order Application,如下所示。

```
package cn.book.order;

import org.springframework.boot.SpringApplication;
import org.springframework.boot.autoconfigure.SpringBootApplication;
import org.springframework.boot.autoconfigure.domain.EntityScan;


@SpringBootApplication
@EntityScan("cn.book.order.entity")
public class OrderApplication {

    public static void main(String[] args) {
        SpringApplication.run(OrderApplication.class, args);
    }
}
```

■ 创建配置文件

创建订单微服务相对应的 YML 配置文件,如下所示。

```yaml
server:
  port: 9003 # 端口
spring:
  application:
    name: order_service # 服务名称
  datasource: # 数据源
    driver-class-name: com.mysql.jdbc.Driver
    url: jdbc:mysql://192.168.10.167:3306/shop?characterEncoding=utf8
    username: root
    password: root
  jpa:
    database: MySQL
    show-sql: true
    open-in-view: true
```

至此,订单微服务创建成功,可以运行 OrderApplication 进行访问。

3.3 使用 Postman 测试微服务

无论是接口调试还是接口测试,Postman 都算得上是很优秀的工具。这里使用 Postman 工具测试微服务的功能,目前这 3 个微服务的功能是相似的,这里选择以商品微服务为例测试其功能。

3.3.1 测试新增

测试新增功能时的操作需要与商品微服务中的控制器新增操作保持一致。

在图 3-6 所示的界面中，选择【POST】请求，输入路由【10.211.55.12:9002/product】，选择发送的请求体【Body】，选择格式【raw】【JSON】，在请求面板中输入 JSON 格式的数据，最后单击【Send】发送请求。获取响应后，在响应面板中查看响应结果。

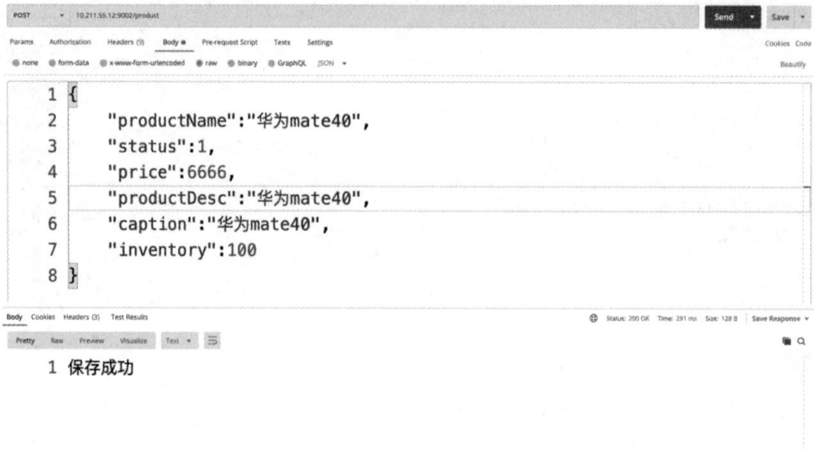

图3-6　新增

3.3.2 测试查询全部

测试查询全部功能时的操作需要与商品微服务中的控制器查询全部操作保持一致。

在图 3-7 所示的界面中，选择【GET】请求，输入路由【10.211.55.12:9002/product】，这里不需要设置参数，直接单击【Send】发送请求。获取响应后，在响应面板中查看响应结果。

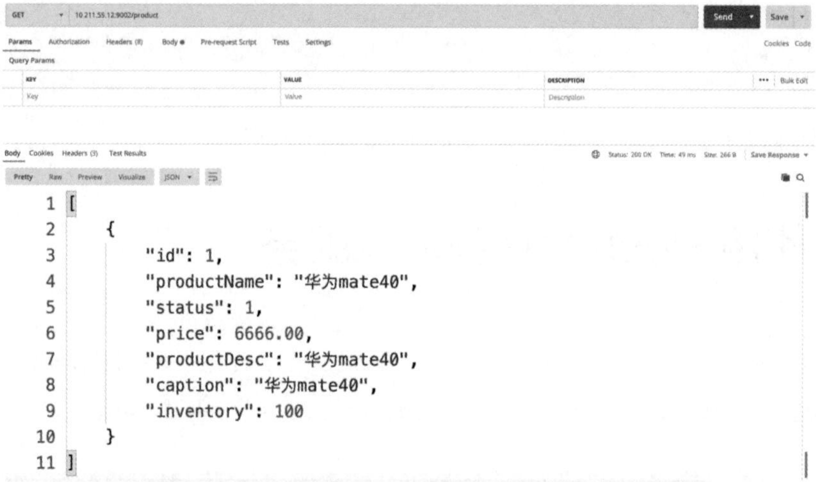

图3-7　查询全部

3.3.3 测试根据 id 查询单个

测试根据 id 查询单个功能时的操作需要与商品微服务中的控制器根据 id 查询单个操作保持一致。

在图 3-8 所示的界面中，选择【GET】请求，输入路由【10.211.55.12:9002/product/1】，其中的 1 是之前新增数据的 id 的值，这里不需要设置参数，直接单击【Send】发送请求，获取响应后，在响应面板中查看响应结果。

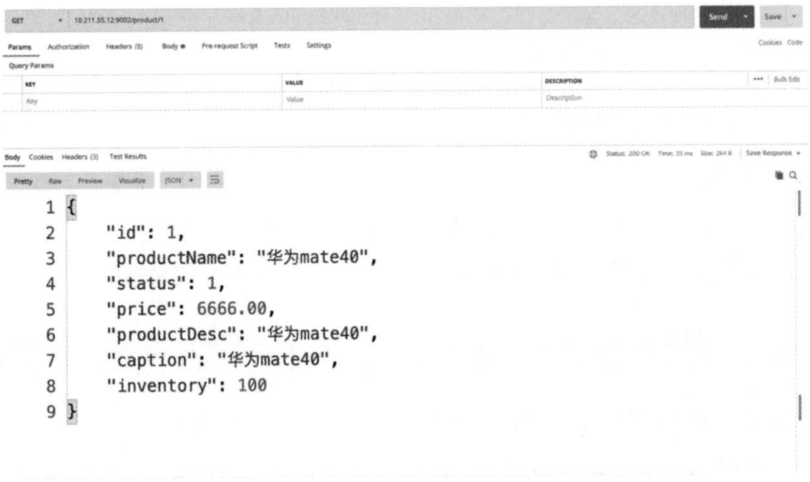

图3-8　根据id查询单个

3.3.4　测试修改

测试修改功能时的操作需要与商品微服务中的控制器修改操作保持一致。

在图 3-9 所示的界面中，选择【PUT】请求，输入路由【10.211.55.12:9002/product/1】，其中的 1 是之前新增数据的 id 的值，选择发送的请求体【Body】，选择格式【raw】、【JSON】，在请求面板中输入 JSON 格式的数据，最后单击【Send】发送请求。获取响应后，在响应面板中查看响应结果。

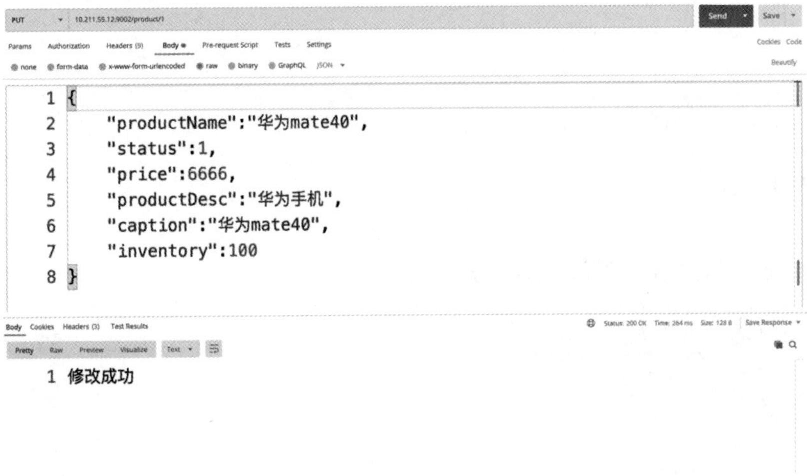

图3-9　修改

再次执行根据 id 查询单个操作进行验证，结果如图 3-10 所示。

图3-10 查询

在图 3-10 所示的界面中发现属性【productDesc】已经修改了，验证修改是成功的。

3.3.5 测试删除

测试删除功能时的操作需要与商品微服务中的控制器删除操作保持一致。

在图 3-11 所示的界面中，选择【DELETE】请求，输入路由【10.211.55.12:9002/product/1】，其中的 1 是之前新增数据的 id 的值，这里不需要设置参数，直接单击【Send】发送请求。获取响应后，在响应面板中查看响应结果。

图3-11 删除

再次执行查询全部操作进行验证，结果如图 3-12 所示。

图3-12 查询全部

在图 3-12 所示的界面中发现属性 id 值是 1 的商品信息已经被删除了，验证删除是成功的。

调用微服务

目前已经编写了 3 个基础的微服务。有这样一个常见场景，在用户下单时需要调用商品微服务获取商品数据。那么应该怎么实现呢？

因为商品微服务提供了可调用的 HTTP 接口，所以可以使用 HTTP 请求的相关工具类来实现下单，如常见的 HttpClient、OkHttp 等，当然也可以使用 Spring 提供的 RestTemplate。

3.4.1 介绍 RestTemplate 类

Spring 框架提供的 RestTemplate 类可用于在应用中调用 REST 服务，它简化了与 HTTP 服务的通信方式，统一了 RESTful 的标准，封装了 HTTP 连接，使用者只需要传入 URL 及返回值类型即可。相较于之前常用的 HttpClient，RestTemplate 是一种更优雅的调用 REST 服务的方式。

在 Spring 应用程序中访问第三方 REST 服务与使用 Spring RestTemplate 类有关。RestTemplate 类的设计原则与许多其他 Spring 模板类（如 JdbcTemplate、JmsTemplate）相同，为执行复杂任务提供了一种具有默认行为的简化方法。

RestTemplate 类默认依赖 JDK 提供 HTTP 连接的能力（HttpURLConnection），如果有需要也可以将 setRequestFactory 方法替换为 Apache HttpComponents、Netty 或 OkHttp 等其他 HTTPlibrary。考虑到 RestTemplate 类是为调用 REST 服务而设计的，因此它的主要方法与 REST 服务的基础方法紧密相连就不足为奇了。前者包括 headForHeaders、getForObject、postForObject、put 和 delete 等方法，后者是 HTTP 的方法（如 HEAD、GET、POST、PUT、DELETE 和 OPTIONS 等）。

3.4.2 使用 RestTemplate 调用微服务

在子工程 order_service 的启动类 OrderApplication 中配置 RestTemplate 对应的 Bean 对象，交给 Spring 进行管理，代码如下所示。

```
@Bean
public RestTemplate restTemplate() {
    return new RestTemplate();
}
```

在项目启动后，RestTemplate 经过初始化，可以通过 @Autowired 注入使用。这里在 OrderController 中模拟一个下单的方法，代码如下所示。

```
@Autowired
private RestTemplate restTemplate;
```

```java
@RequestMapping(value = "/{product_id}/{product_num}", method = RequestMethod.POST)
public String order(@PathVariable Long product_id, @PathVariable Integer product_num) {
    // 定义 URL
    String url = "http://10.211.55.12:9002/product/" + product_id;
    // 使用 restTemplate 调用 HTTP 接口
    Product product = restTemplate.getForEntity(url, Product.class).getBody();
    System.out.printf("\n 下单的商品信息： %s， 数量： %s\n",product,product_num);
    return " 下单成功 ";
}
```

使用 Postman 发送请求，如图 3-13 所示。

图3-13　下单

查看控制台输出的信息，如图 3-14 所示。

下单的商品信息：Product(id=1, productName=华为mate40, status=1, price=6666.00, productDesc=华为mate40, caption=华为mate40, inventory=100)
数量: 2

图3-14　控制台输出的信息

3.4.3　分析硬编码存在的问题

至此，已经可以通过 RestTemplate 调用商品微服务的 RESTful API。但是我们发现服务提供者的网络地址（IP 地址、端口号）等被硬编码到了代码中。这种做法存在许多问题，比如应用场景有局限和无法动态调整，那么怎么解决呢？这就需要通过注册中心动态地进行服务注册和服务发现。

第4章

Eureka 服务注册和发现

Spring Cloud Eureka 是 Spring Cloud Netflix 微服务套件中的一部分,它基于 Netflix Eureka 做了二次封装,主要负责实现微服务架构中的服务治理功能。Spring Cloud 为 Eureka 增加了 Spring Boot 风格的自动化配置,使用者只需通过简单引入依赖和注解配置,就能让 Spring Boot 构建的微服务应用轻松地与 Eureka 服务治理体系进行整合。

本章的主要内容如下。
1. 认识服务注册和服务发现。
2. 使用 Eureka 实现服务注册和发现。
3. Eureka 服务端高可用集群。
4. Eureka 常见问题。
5. Eureka 源码解析。

4.1 认识 Eureka

服务治理可以说是微服务系统中最为核心和基础的模块，它主要用来实现各个微服务实例的自动化注册与发现。为什么在微服务架构中那么需要服务治理模块呢？微服务系统没有它会怎么样呢？

在开始构建微服务系统的时候可能服务并不多，使用者可以通过一些静态配置来完成服务的调用。比如有两个服务 A 和 B，其中服务 A 需要调用服务 B 来完成一个业务操作，为了实现服务 B 的高可用，不论采用服务端负载均衡还是客户端负载均衡，都需要手动维护服务 B 的具体实例清单。随着业务的发展，系统功能越来越复杂，相应的微服务应用也不断增加，静态配置就会变得越来越难以维护。并且面对不断发展的业务，集群规模、服务的位置、服务的命名等都有可能发生变化，如果还是采用手动维护的方式，那么极易发生错误或是出现命名冲突等问题。同时，对于这类静态配置的维护也必将耗费大量的人力。

为了解决微服务架构中的服务实例维护问题，出现了大量的服务治理框架和产品。这些框架和产品的实现都围绕服务注册与服务发现机制来完成对微服务实例的自动化管理。

4.1.1 服务注册和服务发现

服务注册

在服务治理框架中，通常会构建一个注册中心，每个服务单元向注册中心注册自己提供的服务，将主机 IP 地址与端口号、版本号、通信协议等一些附加信息告知注册中心，注册中心按服务名称分类整理服务清单。比如，有两个提供服务 A 的进程分别运行于 192.168.0.100:8000 和 192.168.0.101:8000 位置上，另外还有 3 个提供服务 B 的进程分别运行于 192.168.0.100:9000、192.168.0.101:9000、192.168.0.102:9000 位置上。当这些进程均启动，并向注册中心注册自己的服务之后，注册中心就会维护一个类似表 4-1 的服务清单。

表 4-1 服务清单

服务名称	位置
服务 A	192.168.0.100:8000、192.168.0.101:8000
服务 B	192.168.0.100:9000、192.168.0.101:9000、192.168.0.102:9000

另外，服务注册中心还需要以心跳的方式去监测服务清单中的服务是否可用，若不可用需要将其从服务清单中剔除，达到排除故障服务的目的。

服务发现

由于在服务治理框架下运作，服务间的调用不再通过指定具体的实例位置来实现，而是通过向服务名称发起请求调用来实现，因此，服务消费者在调用服务提供者接口的时候，并不知道具体的服务实例位置。此时，服务消费者需要向注册中心"咨询"服务，并获取所有服务的实例清单，以实现对具体服务实例的访问。比如，现有服务 C 希望调用服务 A，那么服务 C 需要向注册中心发起咨询服务请求，注册中心就会将服务

A 的位置清单返回给服务 C，如按上例服务 A 的情况，服务 C 便获得了服务 A 的两个可用位置 192.168.0.100:8000 和 192.168.0.101:8000。当服务 C 要发起调用的时候，便从该清单中以某种轮询策略取出一个位置来进行服务调用，这就是后续将会介绍的客户端负载均衡。这里只是介绍了一种简单的服务治理逻辑，以便读者理解服务治理框架的基本运行思路。实际的服务治理框架为了确保性能等，基本不会采用每次都向注册中心发起请求以获取服务的方式，并且不同的应用场景在缓存和服务剔除等机制上也会有不同的实现策略。

4.1.2 注册中心

注册中心可以说是微服务架构中的通信录，它记录了服务和服务地址的映射关系。在分布式架构中，服务会注册到这里，当服务需要调用其他服务时，就到这里"找到"服务的地址，从而进行调用。

注册中心是微服务架构中非常重要的一个组件，在微服务架构里主要起到协调者的作用。如图 4-1 所示，注册中心的工作原理主要涉及三大角色：服务提供者、服务消费者、注册中心。

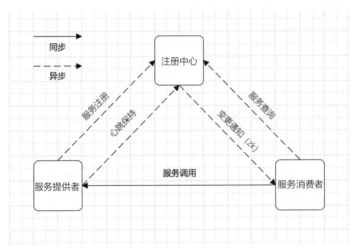

图 4-1 注册中心的工作原理涉及的三大角色

它们之间的关系大致如下。

各个微服务在启动时，将自己的网络地址等信息告知注册中心，注册中心存储这些数据。

服务消费者从注册中心查询服务提供者的地址，并通过该地址调用服务提供者的接口。

各个微服务与注册中心使用一定的机制（如心跳）通信。如果注册中心与某个微服务长时间无法通信，就会注销该微服务的实例。

微服务网络地址发生变化（如实例增加或 IP 地址变动等）时，会重新告知注册中心。这样，服务消费者就无须人工修改服务提供者的网络地址了。

注册中心应具备以下功能。

提供服务注册表。服务注册表是注册中心的核心，用来记录各个微服务的信息，例如微服务的名称、IP 地址、端口号等。服务注册表提供查询 API 和管理 API，查询 API 用于查询可用的微服务实例，管理

API 用于服务的注册与注销。

服务注册与发现。服务注册是指微服务在启动时，将自己的信息告知注册中心的过程。服务发现是指查询可用的微服务列表及网络地址的机制。

服务检查。注册中心使用一定的机制定时检查已注册的服务，若发现某服务实例长时间无法访问，就会从服务注册表中移除该实例。

互联网架构下，大部分系统已经转型为分布式系统。其中，注册中心是分布式系统中非常重要的组成部分。如今按需选择合适的注册中心，变得尤为重要。下面介绍常见的注册中心。

■ Eureka

Eureka 是 Spring Cloud "全家桶"中非常重要的一个组件，主要实现服务的注册和发现。Eureka 做到了 CAP 理论［分布式数据库中的一种弱一致性理论，即系统最多只能同时满足一致性（Consistency）、可用性（Availability）、容忍网络分割（Partition Tolerance）等 3 个需求中的 2 个］中的 AP，强调服务的高可用性，实现中分 Eureka 服务端和 Eureka 客户端两部分。

Eureka 客户端会向 Eureka 注册中心注册服务，并通过心跳来更新它的服务租约。同时 Eureka 客户端可以从 Eureka 服务端查询当前注册的服务信息，把它们缓存到本地并周期性地刷新服务状态。若服务集群出现分区故障，Eureka 会进入自动保护模式，允许分区故障的节点继续提供服务；若分区故障恢复，集群中的其他分区会再次同步它们的状态。

Spring Cloud 对 Eureka 做了非常好的集成封装，是官方推荐的注册中心。

■ ZooKeeper

ZooKeeper 是大数据 Hadoop 中的一个分布式调度组件，强调数据一致性和扩展性，可用于服务的注册和发现。它是 Dubbo 中默认的注册中心，也是目前使用非常广泛的分布式服务发现组件。ZooKeeper 注重 CAP 理论中的 CP。

■ Consul

Consul 是一个高可用的分布式注册中心，由 HashiCorp 公司推出，是使用 Go 语言实现的开源、共享的服务工具。Consul 在分布式服务注册与发现方面有自己的特色，解决方案更加"一站式"，不再需要依赖其他工具。

Consul 通过 HTTP 接口和 DNS 协议调用 API 存储键值对，使服务注册和服务发现更容易；支持健康检查，可以快速地告警在集群中的操作；支持键值对存储动态配置；支持任意数量的区域。

■ Nacos

Nacos 是一个更易于构建云原生应用的动态服务发现、配置管理和服务管理平台。简单来说，Nacos 就是注册中心和配置中心的组合，提供简单易用的特性集，帮助使用者解决微服务开发过程中必会涉及的服务注册与发现、服务配置、服务管理等问题。Nacos 还是 Spring Cloud Alibaba 组件之一，负责服务注册与发现。

我们可通过表 4-2 了解 Eureka、Nacos、Consul、ZooKeeper 的异同点。选择什么类型的服务注册与发现组件可以根据自身项目要求决定。

表 4-2 注册中心对比

注册中心 特性	Eureka	Nacos	Consul	ZooKeeper
CAP	AP	CP+AP	CP	CP
健康检查	Client Beat	TCP/HTTP/MySQL/Client Beat	TCP/HTTP/grpc/cmd	Keep Alive
雪崩保护	有	有	无	无
自动注销实例	支持	支持	不支持	支持
访问协议	HTTP	HTTP/DNS	HTTP/DNS	TCP
监听协议	支持	支持	支持	支持
多数据中心	支持	支持	支持	不支持
跨注册中心同步	不支持	支持	支持	不支持
Spring Cloud 集成	支持	支持	支持	支持

4.1.3 Eureka 框架的原理

Eureka 是 Netflix 开发的服务发现框架，Spring Cloud 将它集成在自己的子项目 Spring Cloud Netflix 中，以实现 Spring Cloud 的服务发现功能。Eureka 的基本架构由 3 个角色组成：提供服务注册和发现 Eureka 服务端；服务提供者（Service Provider）将自身服务注册到 Eureka，从而使服务消费者能够找到；服务消费者（Service Consumer）从 Eureka 获取注册服务列表，从而能够消费服务。Eureka 的架构如图 4-2 所示。

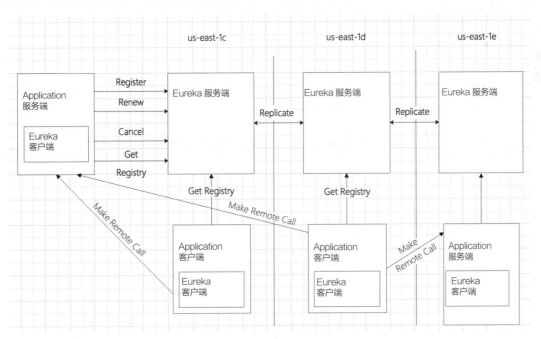

图 4-2 Eureka 架构

图 4-2 所示的架构来自 Eureka 官网，大致描述了 Eureka 的工作过程。图中包含的组件非常多，可能比较难以理解，下面用通俗易懂的语言解释一下。

Application 服务端相当于本书中的服务提供者，Application 客户端相当于服务消费者；Make Remote Call 可以简单理解为调用 RESTful API；us-east-1c、us-east-1d 等都是 Zone，它们都属于 us-east-1 这个 Region。

由图 4-2 可知，Eureka 包含两个组件：Eureka 服务端和 Eureka 客户端，它们的作用如下。

Eureka 客户端是一个 Java 客户端，用于简化与 Eureka 服务端的交互。

Eureka 服务端提供服务发现的能力，各个微服务启动时，会通过 Eureka 客户端向 Eureka 服务端注册自己的信息（如网络信息），Eureka 服务端会存储该服务的信息。

微服务启动后，会周期性地向 Eureka 服务端发送心跳（默认周期为 30s）以续约自己的信息。如果 Eureka 服务端在一定时间（默认为 90s）内没有接收到某个微服务节点的心跳，将会注销该微服务节点。

每个 Eureka 服务端同时也是 Eureka 客户端，多个 Eureka 服务端之间通过复制（Replicate）的方式完成服务注册表的同步。

Eureka 客户端会缓存 Eureka 服务端中的信息。即使所有的 Eureka 服务端节点都宕掉，服务消费者依然可以使用缓存中的信息找到服务提供者。

总结一下，Eureka 通过心跳检测、健康检查和客户端缓存等机制，提高了系统的灵活性、可伸缩性和可用性。

4.2 使用 Eureka

现在我们基本了解了 Eureka 的基本原理和架构。在贯穿案例中已经搭建了 3 个微服务，下面通过 Eureka 搭建注册中心，实现微服务的注册和发现。

4.2.1 搭建 Eureka 注册中心

■ 创建工程

使用 IntelliJ IDEA 创建子工程——Eureka 注册中心（eureka_server），如图 4-3 所示。

图 4-3 注册中心

创建子工程后,在其对应的 pom.xml 文件中导入 Maven 坐标,如下所示。

```
<dependency>
    <groupId>org.springframework.cloud</groupId>
    <artifactId>spring-cloud-starter-netflix-eureka-server</artifactId>
</dependency>
```

创建启动类

创建注册中心相对应的启动类 EurekaServerApplication,通过 @EnableEurekaServer 注解启动一个注册中心提供给其他应用进行对话。这一步非常简单,只需在一个普通的 Spring Boot 应用中添加这个注解就能启用此功能,如下所示。

```
package cn.book.eureka;

import org.springframework.boot.SpringApplication;
import org.springframework.boot.autoconfigure.SpringBootApplication;
import org.springframework.boot.autoconfigure.domain.EntityScan;
import org.springframework.cloud.netflix.eureka.server.EnableEurekaServer;
import org.springframework.context.annotation.Bean;
import org.springframework.web.client.RestTemplate;

@SpringBootApplication
```

```
// 激活 Eureka 服务端
@EnableEureka Server
public class EurekaServerApplication {

    public static void main(String[] args) {
        SpringApplication.run(EurekaServerApplication.class, args);
    }
}
```

■ 创建配置文件

创建注册中心相对应的 YML 配置文件，如下所示。

```
server:
  port: 9000 # 端口
spring:
  application:
    name: eureka_server # 服务名称
eureka: # 配置 Eureka 服务端
  instance:
    prefer-ip-address: true
  client:
    register-with-eureka: false # 是否将自己注册到注册中心
    fetch-registry: false # 是否从 Eureka 中获取注册信息
    service-url: # 配置暴露给 Eureka 客户端的请求地址
      defaultZone: http://10.211.55.12:9000/eureka/
  server:
    enable-self-preservation: false # 关闭自我保护
    eviction-interval-timer-in-ms: 1000 # 剔除服务间隔
```

■ 访问注册中心管理后台

在完成了上面的配置后，启动应用并访问 http://10.211.55.12:9000/eureka_server/，可以看到图 4-4 所示的窗口。

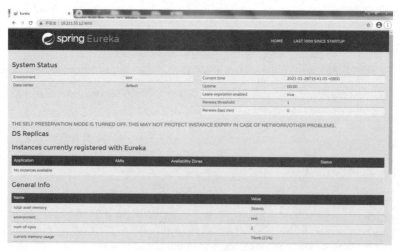

图 4-4　访问注册中心管理后台

在图 4-4 中 Instances currently registered with Eureka 栏是空的，说明该注册中心还没有注册任何服务。

4.2.2 将服务注册到 Eureka 注册中心

■ 导入 Maven 坐标

在商品微服务的 pom.xml 文件中导入 Maven 坐标，如下所示。

```
<dependency>
    <groupId>org.springframework.cloud</groupId>
    <artifactId>spring-cloud-starter-netflix-eureka-client</artifactId>
</dependency>
```

■ 修改配置文件

在工程的配置文件 application.yml 中添加 Eureka 服务端的主机地址，如下所示。

```
eureka: # 配置 Eureka
  client:
    service-url:
      defaultZone: http://10.211.55.12:9000/eureka/ # 多个 Eureka 服务端之间用逗号隔开
  instance:
    prefer-ip-address: true # 使用 IP 地址注册
```

■ 再次访问注册中心管理后台

设置用户微服务和订单微服务。

重新启动 Eureka 注册中心和商品微服务，再次访问注册中心管理后台，可以看到图 4-5 所示的窗口。

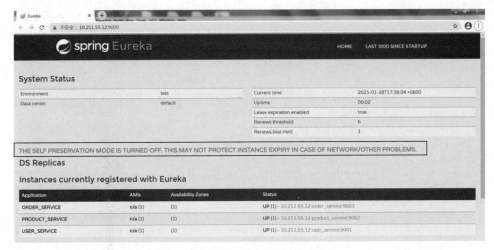

图 4-5 访问注册中心管理后台

在图 4-5 中 Instances currently registered with Eureka 栏下已经出现注册的 3 个微服务。

那图中框住的内容是什么意思呢？

默认情况下，如果 Eureka 服务端在一定时间（默认为 90s）内没有接收到某个微服务实例的心跳，那么 Eureka 服务端会注销该实例。但是当发生网络分区故障时，微服务与 Eureka 服务端之间无法正常通信，以上行为可能变得非常危险，因为微服务本身是健康的，不应该注销这个微服务。Eureka 通过"自我保护模式"来解决这个问题。当 Eureka 服务端节点在短时间内丢失过多 Eureka 客户端时（可能发生了网络分区故障），这个节点就会进入自我保护模式。一旦进入该模式，Eureka 服务端就会保护服务注册表中的信息，不再删除服务注册表中的数据（也就是不会注销任何微服务）。当网络分区故障恢复后，该 Eureka 服务端节点会自动退出自我保护模式。Eureka 注册中心可以通过设置 eureka.enableSelfPreservation=false 来退出自我保护模式。

4.2.3 使用 Eureka 的元数据完成服务调用

Eureka 的元数据有两种：标准元数据和自定义元数据。

标准元数据包括主机名、IP 地址、端口号、状态页和健康检查等信息，这些信息都会被发布在服务注册表中，用于服务之间的调用。

自定义元数据可以使用 eureka.instance.metadata-map 配置符合键值对的存储格式。自定义元数据可以在远程客户端中访问。

在程序中可以使用 DiscoveryClient 获取指定微服务的所有元数据信息来改写 restTemplate 调用的 url，从而解决硬编码的问题，从而代码如下所示。

```java
@Autowired
 private DiscoveryClient discoveryClient;

@RequestMapping(value = "/{product_id}/{product_num}", method = RequestMethod.GET)
 public Map<String, Object> order(@PathVariable Long product_id, @PathVariable Integer product_num) {
    // 通过指定微服务的名称获取该微服务的元数据
    ServiceInstance product_service = discoveryClient.getInstances("product_service").get(0);
    // 定义 URL
        String url = String.format("http://%s:%s/product/%d", product_service.getHost(), product_service.getPort(), product_id);
    // 使用 restTemplate 调用 HTTP 接口
    Product product = restTemplate.getForEntity(url, Product.class).getBody();
    // 创建字典对象，存放元数据信息
    Map<String, Object> map = new HashMap<String, Object>();
    // 存数据
    map.put("product_num", product_num);
    map.put("product", product);
    // 返回数据
    return map;
 }
```

上面的代码使用注入的方式获取 DiscoveryClient 对象，接着使用 getInstances 方法根据微服务的名称获取该微服务对象以及相应的微服务信息，比如 IP 地址和端口号，将信息拼接到 url 中，这样就解决了微服务调用 url 硬编码的问题。

重启服务后，访问路由，结果如图 4-6 所示。

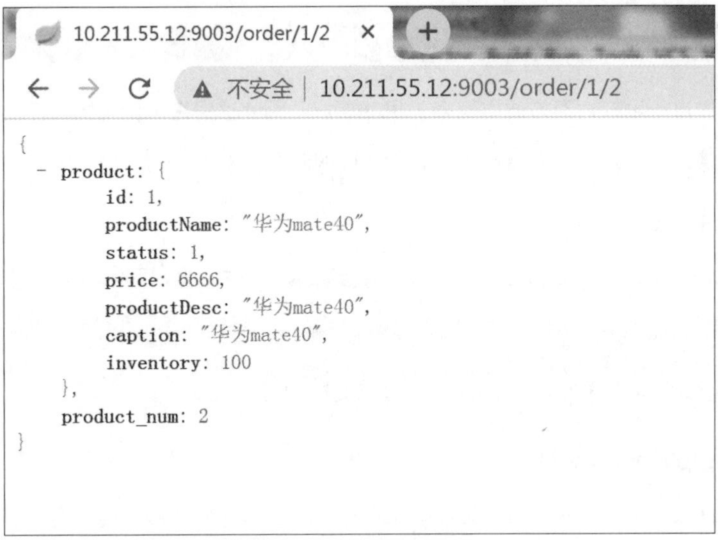

图 4-6 访问结果

4.3 Eureka 服务端高可用集群

目前实现了单节点的 Eureka 服务端的服务注册与服务发现功能。Eureka 客户端会定时连接 Eureka 服务端，以获取注册表中的信息并缓存到本地。微服务在消费远程 API 时总是使用本地缓存中的数据。因此一般来说，即使 Eureka 服务端发生宕机，也不会影响到服务之间的调用。但如果 Eureka 服务端宕机时，某些微服务也出现了不可用的情况，Eureka 服务端中的缓存若不被刷新，就可能会影响到微服务的调用，甚至影响到整个应用系统的高可用性。因此，在生成环境中，通常会部署一个高可用的 Eureka 服务端集群，如图 4-7 所示。

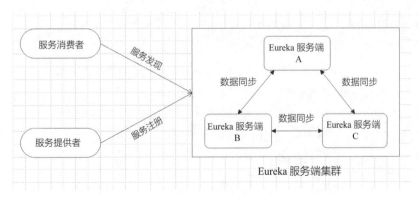

图 4-7 Eureka服务端集群

065

Eureka 服务端可以通过运行多个实例并以相互注册的方式实现高可用部署，Eureka 服务端实例彼此信息增量同步，从而确保所有节点的数据一致。事实上，节点之间相互注册是 Eureka 服务端的默认行为。

4.3.1 搭建 Eureka 服务端高可用集群

由于是在 PC 中进行测试，不方便模拟多主机的情况，因此这里在一台机器上启动 3 个端口号不同的 Eureka 服务端进行模拟，端口号分别是 8001、8002 和 8003。

■ 修改 hosts 文件

通过路径 C:\Windows\System32\drivers\etc\hosts 找到 hosts 文件，将如下代码添加到 hosts 文件最后。

```
127.0.0.1    eureka8001.com
127.0.0.1    eureka8002.com
127.0.0.1    eureka8003.com
```

■ 修改配置文件 application.yml

这里主要添加 hostname、修改 defaultZone。defaultZone 的值为其他 Eureka 服务端的地址，多个地址之间使用逗号隔开。

修改端口号为 8001 的 Eureka 服务端的配置文件 application.yml，修改内容如下所示。

```yaml
server:
  port: 8001 # 端口
spring:
  application:
    name: eureka_server_8001 # 服务名称
eureka: # 配置 Eureka 服务端
  instance:
    prefer-ip-address: true
    hostname: eureka8001.com # 主机名
  client:
    register-with-eureka: false # 不将自己注册到注册中心
    fetch-registry: false # 不从 Eureka 中获取注册信息
    service-url: # 配置暴露给 Eureka 客户端的请求地址
      defaultZone: http://eureka8002.com:8002/eureka/,http://eureka8003.com:8003/eureka/
  server:
    enable-self-preservation: false # 关闭自我保护
    eviction-interval-timer-in-ms: 1000 # 剔除服务间隔
```

修改端口号为 8002 的 Eureka 服务端的配置文件 application.yml，修改内容如下所示。

```yaml
server:
  port: 8002 # 端口
spring:
  application:
    name: eureka_server_8002 # 服务名称
eureka: # 配置 Eureka 服务端
  instance:
```

```
    prefer-ip-address: true
    hostname: eureka8002.com # 主机名称
  client:
    register-with-eureka: false # 不将自己注册到注册中心
    fetch-registry: false # 不从 Eureka 中获取注册信息
    service-url: # 配置暴露给 Eureka 客户端的请求地址
      defaultZone: http://eureka8001.com:8001/eureka/,http://eureka8003.com:8003/eureka/
  server:
    enable-self-preservation: false # 关闭自我保护
    eviction-interval-timer-in-ms: 1000 # 剔除服务间隔
```

修改端口号为 8003 的 Eureka 服务端的配置文件 application.yml，修改内容如下所示。

```
server:
  port: 8003 # 端口
spring:
  application:
    name: eureka_server_8003 # 服务名称
eureka: # 配置 Eureka 服务端
  instance:
    prefer-ip-address: true
    hostname: eureka8003.com # 主机名称
  client:
    register-with-eureka: false # 不将自己注册到注册中心
    fetch-registry: false # 不从 Eureka 中获取注册信息
    service-url: # 配置暴露给 Eureka 客户端的请求地址
      defaultZone: http://eureka8001.com:8001/eureka/,http://eureka8002.com:8002/eureka/
  server:
    enable-self-preservation: false # 关闭自我保护
    eviction-interval-timer-in-ms: 1000 # 剔除服务间隔
```

■ 访问注册中心管理后台

分别启动端口号为 8001、8002 和 8003 的 Eureka 服务端主机，分别访问注册中心管理后台，可以看到图 4-8、图 4-9 和图 4-10 所示的窗口。

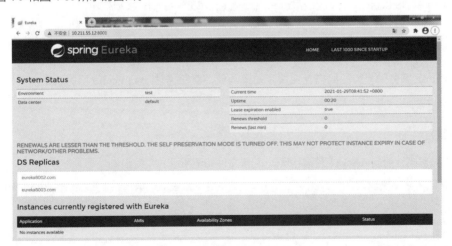

图 4-8　端口号为8001的Eureka服务端主机

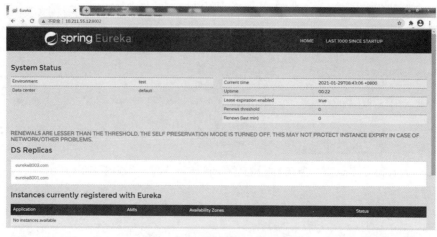

图4-9　端口号为8002的Eureka服务端主机

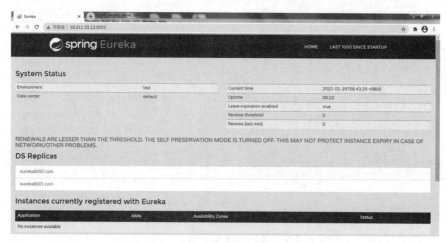

图4-10　端口号为8003的Eureka服务端主机

在图4-10中DS Replicas栏下已经显示Eureka服务端集群中的其他主机信息。

4.3.2　将服务注册到 Eureka 服务端集群

目前Eureka服务端集群已经搭建完成，如果需要将微服务注册到Eureka服务端集群，只需要修改YML配置文件即可。

■ 注册微服务

将商品微服务注册到Eureka服务端集群，配置文件如下所示。

```
server:
  port: 9002 # 端口
spring:
  application:
    name: product_service # 服务名称
  datasource: # 数据源
    driver-class-name: com.mysql.jdbc.Driver
```

```
      url: jdbc:mysql://192.168.10.167:3306/shop?characterEncoding=utf8
      username: root
      password: root
    jpa:
      database: MySQL
      show-sql: true
      open-in-view: true
eureka: # 配置 Eureka
  client:
    service-url:
      defaultZone: http://eureka8001.com:8001/eureka/,http://eureka8002.com:8002/eureka/,http://eureka8003.com:8003/eureka/ # 多个 Eureka 服务端之间用逗号隔开
  instance:
    prefer-ip-address: true # 使用 IP 地址注册
```

defaultZone 配置了集群中所有主机的地址，当然只配置一个主机地址也是可以的，集群可以将微服务的数据进行同步，但是为了防止某个主机宕机，还是建议 defaultZone 配置集群中所有主机的地址，多个地址之间使用逗号隔开。

■ 访问注册中心管理后台

配置完成后，分别启动端口号为 8001、8002 和 8003 的 Eureka 服务端主机，分别访问注册中心管理后台，可以看到图 4-11、图 4-12 和图 4-13 所示的窗口。

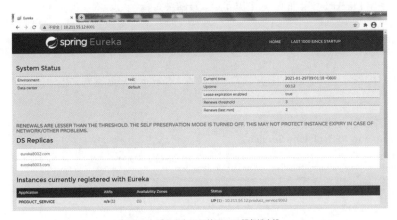

图 4-11　端口号为8001的Eureka服务端主机

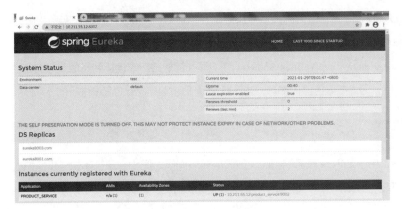

图 4-12　端口号为8002的Eureka服务端主机

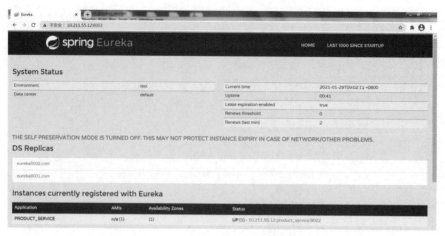

图 4-13　端口号为8003的Eureka服务端主机

在图 4-13 中 Instances currently registered with Eureka 栏下已经显示微服务相关信息。

Eureka 常见问题

下面介绍在使用 Eureka 时遇到的一些常见问题以及解决方案。

4.4.1　服务注册慢

默认情况下，服务注册到 Eureka 服务端的过程较慢。Spring Cloud 官方文档中给出了详细的解释，如图 4-14 所示。

1.10 Why Is It so Slow to Register a Service?

Being an instance also involves a periodic heartbeat to the registry (through the client's `serviceUrl`) with a default duration of 30 seconds. A service is not available for discovery by clients until the instance, the server, and the client all have the same metadata in their local cache (so it could take 3 heartbeats). You can change the period by setting `eureka.instance.leaseRenewalIntervalInSeconds`. Setting it to a value of less than 30 speeds up the process of getting clients connected to other services. In production, it is probably better to stick with the default, because of internal computations in the server that make assumptions about the lease renewal period.

图 4-14　Spring Cloud官方文档说明

翻译之后的大致意思是：服务的注册涉及心跳，默认心跳间隔为 30s。在实例、服务器、客户端都在本地缓存中具有相同的元数据之前，服务不可用于客户端发现（所以可能需要 3 次心跳）。可以通过配置 eureka.instance.leaseRenewalIntervalInSeconds（心跳频率）加快客户端连接到其他服务的过程。在生产中，建议坚持使用默认值，因为在服务器内部有一些计算是基于对续约做出的假设进行的。

4.4.2　服务节点剔除问题

默认情况下，由于 Eureka 服务端剔除失效服务节点间隔时间为 90s 且存在自我保护的机制，因此不能有效且迅速地剔除失效服务节点，这会给开发或测试造成困扰。解决方案如下所示。

■ **Eureka** 服务端配置关闭自我保护，并设置剔除失效节点的时间间隔

```
server:
  enable-self-preservation: false # 关闭自我保护
  eviction-interval-timer-in-ms: 1000 # 剔除失效节点的时间间隔
```

■ **Eureka** 客户端配置开启健康检查，并设置续约时间

```
eureka:
  client:
    healthcheck: true # 开启健康检查（依赖 spring-boot-actuator）
    serviceUrl:
      defaultZone: http://eureka8001.com:8001/eureka/
  instance:
    prefer-ip-address: true
    lease-expiration-duration-in-seconds: 10 #Eureka 客户端发送心跳给 Eureka 服务端后，
续约到期时间（默认为 90s）
    lease-renewal-interval-in-seconds: 5 # 发送心跳续约间隔（默认为 30s）
```

在开发过程中，建议将续约和剔除时间设置为较小的数字，这样可以缩短等待时间。

4.4.3 监控页面显示 IP 地址信息

在 Eureka 服务端的注册中心管理后台中，默认情况下显示的服务实例名称是由微服务定义的名称和端口号组成的。为了更好地对所有服务进行定位，微服务注册到 Eureka 服务端的时候可以手动配置示例 id。配置方式如下所示。

```
eureka: # 配置 Eureka
  client:
    service-url:
      defaultZone: http://eureka8001.com:8001/eureka/,http://eureka8002.com:8002/eureka/,http://eureka8003.com:8003/eureka/ # 多个 Eureka 服务端之间用逗号隔开
  instance:
    prefer-ip-address: true # 使用 IP 地址注册
    instance-id: ${spring.cloud.client.ip-address}:${server.port} # 向注册中心注册服务 id
```

重启微服务，刷新注册中心管理后台，可以看到监控页面显示微服务的 IP 地址信息，如图 4-15 所示。

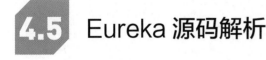

图 4-15　显示 IP 地址信息

4.5　Eureka 源码解析

Eureka 服务端作为一个开箱即用的注册中心，提供以下 6 个功能，以满足与 Eureka 客户端交互的需求。

（1）服务注册。

（2）接收服务心跳。

（3）服务剔除。

（4）服务下线。

（5）集群同步。

（6）获取注册表中服务实例的信息。

需要注意的是，Eureka 服务端同时也是 Eureka 客户端，在不禁止 Eureka 服务端的客户端行为时，它会向它的配置文件中的其他 Eureka 服务端进行拉取注册表、注册服务和发送心跳等操作。

4.5.1 服务注册表

InstanceRegistry 是 Eureka 服务端中管理注册表的接口。它的类图如图 4-16 所示。图 4-16 中出现了两个 InstanceRegistry，最下面的 InstanceRegistry 对 Eureka 服务端的注册表实现类 PeerAwareInstanceRegistryImpl 进行了继承和扩展，使其适配 Spring Cloud 的使用环境，主要实现由 PeerAwareInstanceRegistryImpl 提供。上面的 InstanceRegistry 是 Eureka 服务端中管理注册表的核心接口，其职责是在内存中管理注册到 Eureka 服务端中的服务实例信息。InstanceRegistry 接口分别继承了 LeaseManager 接口和 LookupService 接口。LeaseManager 接口的功能是对注册到 Eureka 服务端中的服务实例租约进行管理，而 LookupService 接口提供对服务实例进行检索的功能。

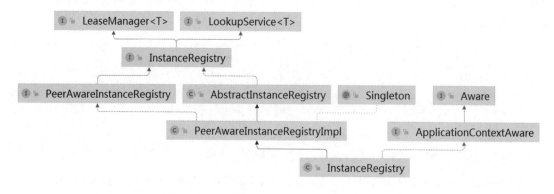

图 4-16 InstanceRegistry 类图

LeaseManager 接口提供的方法如下所示。

```
public interface LeaseManager<T> {
    void register(T var1, int var2, boolean var3);

    boolean cancel(String var1, String var2, boolean var3);

    boolean renew(String var1, String var2, boolean var3);

    void evict();
}
```

LeaseManager 接口的作用是对注册到 Eureka 服务端中的服务实例租约进行管理，包括服务注册、服务下线、服务租约更新以及服务剔除等操作。

LeaseManager 中管理的对象是 Lease，Lease 代表一个 Eureka 客户端服务实例信息的租约，它提供了对其内持有的类的时间有效性操作。Lease 持有的类是代表服务实例信息的 InstanceInfo。Lease 中定义了租约的操作类型，包括注册、下线、更新等，同时提供了对租约中时间属性的各项操作。租约默认有效时长（duration）为 90s。

InstanceRegistry 接口在继承 LeaseManager 接口和 LookupService 接口的基础上，添加了一些特有的方法，可以更为简单地管理服务实例租约和查询注册表中的服务实例信息。可以通过 AbstractInstanceRegistry 来查看 InstanceRegistry 接口方法的具体实现。

PeerAwareInstanceRegistry 接口在继承 InstanceRegistry 接口的基础上，添加了 Eureka 服务端集群同步的方法，其实现类 PeerAwareInstanceRegistryImpl 继承了 AbstractInstanceRegistry 类的实现，在对本地注册表操作的基础上添加了对其 Peer 节点进行同步复制操作，使得 Eureka 服务端集群中的注册表信息能够保持一致。

4.5.2 服务注册

Eureka 客户端在发起服务注册时会将自身的服务实例元数据封装在 InstanceInfo 中，然后将 InstanceInfo 发送到 Eureka 服务端。Eureka 服务端在接收到 Eureka 客户端发送的 InstanceInfo 后，会尝试将其放到本地注册表中，以供其他 Eureka 客户端进行服务发现。

服务注册的主要实现在 AbstractInstanceRegistry#register 方法中，代码如下所示。

```java
public void register(InstanceInfo registrant, int leaseDuration, boolean isReplication) {
    try {
        this.read.lock();
        Map<String, Lease<InstanceInfo>> gMap = (Map)this.registry.get(registrant.getAppName());
        EurekaMonitors.REGISTER.increment(isReplication);
        if (gMap == null) {
            ConcurrentHashMap<String, Lease<InstanceInfo>> gNewMap = new ConcurrentHashMap();
            gMap = (Map)this.registry.putIfAbsent(registrant.getAppName(), gNewMap);
            if (gMap == null) {
                gMap = gNewMap;
            }
        }
        Lease<InstanceInfo> existingLease = (Lease)((Map)gMap).get(registrant.getId());
        if (existingLease != null && existingLease.getHolder() != null) {
            Long existingLastDirtyTimestamp = ((InstanceInfo)existingLease.getHolder()).getLastDirtyTimestamp();
            Long registrationLastDirtyTimestamp = registrant.getLastDirtyTimestamp();
            logger.debug("Existing lease found (existing={}, provided={}", existingLastDirtyTimestamp, registrationLastDirtyTimestamp);
```

```java
            if (existingLastDirtyTimestamp > registrationLastDirtyTimestamp) {
                logger.warn("There is an existing lease and the existing lease's dirty timestamp {} is greater than the one that is being registered {}", existingLastDirtyTimestamp, registrationLastDirtyTimestamp);
                logger.warn("Using the existing instanceInfo instead of the new instanceInfo as the registrant");
                registrant = (InstanceInfo)existingLease.getHolder();
            }
        } else {
            synchronized(this.lock) {
                if (this.expectedNumberOfClientsSendingRenews > 0) {
                    ++this.expectedNumberOfClientsSendingRenews;
                    this.updateRenewsPerMinThreshold();
                }
            }

            logger.debug("No previous lease information found; it is new registration");
        }

        Lease<InstanceInfo> lease = new Lease(registrant, leaseDuration);
        if (existingLease != null) {
            lease.setServiceUpTimestamp(existingLease.getServiceUpTimestamp());
        }

        ((Map)gMap).put(registrant.getId(), lease);
        synchronized(this.recentRegisteredQueue) {
            this.recentRegisteredQueue.add(new Pair(System.currentTimeMillis(), registrant.getAppName() + "(" + registrant.getId() + ")"));
        }

        if (!InstanceStatus.UNKNOWN.equals(registrant.getOverriddenStatus())) {
            logger.debug("Found overridden status {} for instance {}. Checking to see if needs to be add to the overrides", registrant.getOverriddenStatus(), registrant.getId());
            if (!this.overriddenInstanceStatusMap.containsKey(registrant.getId())) {
                logger.info("Not found overridden id {} and hence adding it", registrant.getId());
                this.overriddenInstanceStatusMap.put(registrant.getId(), registrant.getOverriddenStatus());
            }
        }

        InstanceStatus overriddenStatusFromMap = (InstanceStatus)this.overriddenInstanceStatusMap.get(registrant.getId());
        if (overriddenStatusFromMap != null) {
            logger.info("Storing overridden status {} from map", overriddenStatusFromMap);
            registrant.setOverriddenStatus(overriddenStatusFromMap);
        }

        InstanceStatus overriddenInstanceStatus = this.getOverriddenInstanceStatus(registrant, existingLease, isReplication);
        registrant.setStatusWithoutDirty(overriddenInstanceStatus);
        if (InstanceStatus.UP.equals(registrant.getStatus())) {
            lease.serviceUp();
        }

        registrant.setActionType(ActionType.ADDED);
```

```
            this.recentlyChangedQueue.add(new AbstractInstanceRegistry.
RecentlyChangedItem(lease));
        registrant.setLastUpdatedTimestamp();
            this.invalidateCache(registrant.getAppName(), registrant.getVIPAddress(), registrant.
getSecureVipAddress());
        logger.info("Registered instance {}/{} with status {} (replication={})", new Object[]{registrant.
getAppName(), registrant.getId(), registrant.getStatus(), isReplication});
    } finally {
        this.read.unlock();
    }

}
```

在 register 方法中，服务实例的 InstanceInfo 保存在 Lease 中，Lease 在 AbstractInstanceRegistry 中统一通过 ConcurrentHashMap 保存在内存中。在服务注册过程中，会先获取一个读锁，防止其他线程对注册表进行数据操作，避免数据的不一致。然后查询对应的租约是否已经存在于注册表中，注册表根据 appName 划分服务集群，使用 InstanceId 唯一标记服务实例。如果租约存在，比较两个租约中的 InstanceInfo 的 LastUpdatedTimestamp（最后更新时间），保留时间戳大的 InstanceInfo。如果租约不存在，意味着这是一次全新的服务注册，将会进行自我保护的统计，创建新的租约保存 InstanceInfo。接着将租约放到注册表中。

之后将进行一系列缓存操作并根据覆盖状态规则设置服务实例的状态，缓存操作包括将 InstanceInfo 分别加入用于统计 Eureka 客户端增量式获取注册表信息的 recentlyChangedQueue 和失效 invalidateCache 中对应的缓存。最后设置服务实例租约的上线时间用于计算租约的有效时间，释放读锁并完成服务注册。

register 方法中有诸多的同步操作，以防止数据被错误地覆盖，有兴趣的读者可以细细研究一下，在此不展开讲述。

4.5.3 接收服务心跳

Eureka 客户端完成服务注册之后，需要定时向 Eureka 服务端发送心跳请求（默认 30s 一次），以维持自己在 Eureka 服务端中租约的有效性。

Eureka 服务端处理心跳请求的核心逻辑在 AbstractInstanceRegistry#renew 方法中。renew 方法用于对 Eureka 客户端位于注册表中的租约完成续约操作。不像 register 方法需要服务实例信息，renew 方法仅根据服务实例的服务名称和服务实例 id 即可更新对应租约的有效时间。具体代码如下所示。

```
public boolean renew(String appName, String id, boolean isReplication) {
    EurekaMonitors.RENEW.increment(isReplication);
    Map<String, Lease<InstanceInfo>> gMap = (Map)this.registry.get(appName);
    Lease<InstanceInfo> leaseToRenew = null;
    if (gMap != null) {
        leaseToRenew = (Lease)gMap.get(id);
    }

    if (leaseToRenew == null) {
        EurekaMonitors.RENEW_NOT_FOUND.increment(isReplication);
```

```
            logger.warn("DS: Registry: lease doesn't exist, registering resource: {} - {}", appName,
id);
            return false;
        } else {
            InstanceInfo instanceInfo = (InstanceInfo)leaseToRenew.getHolder();
            if (instanceInfo != null) {
                InstanceStatus overriddenInstanceStatus = this.getOverriddenInstanceStatus(instanceIn
fo, leaseToRenew, isReplication);
                if (overriddenInstanceStatus == InstanceStatus.UNKNOWN) {
                    logger.info("Instance status UNKNOWN possibly due to deleted override for instance {};
re-register required", instanceInfo.getId());
                    EurekaMonitors.RENEW_NOT_FOUND.increment(isReplication);
                    return false;
                }

                if (!instanceInfo.getStatus().equals(overriddenInstanceStatus)) {
                    logger.info("The instance status {} is different from overridden instance status {} for
instance {}. Hence setting the status to overridden status", new Object[]{instanceInfo.getStatus().
name(), instanceInfo.getOverriddenStatus().name(), instanceInfo.getId()});
                    instanceInfo.setStatusWithoutDirty(overriddenInstanceStatus);
                }
            }

            this.renewsLastMin.increment();
            leaseToRenew.renew();
            return true;
        }
    }
```

在 renew 方法中,不关注 InstanceInfo,仅关注租约本身以及租约的服务实例状态。如果根据服务实例的 appName 和 InstanceInfoId 查询出服务实例的租约,并且根据 getOverriddenInstanceStatus 方法得到的 InstanceStatus 不为 InstanceStatus.UNKNOWN,那么更新租约中的有效时间,即更新租约 Lease 中的 LastUpdateTimestamp,以达到续约的目的;如果租约不存在,那么返回续约失败。

4.5.4 服务剔除

如果 Eureka 客户端在注册后,既没有续约,也没有下线(服务崩溃或者网络异常等原因),那么服务就处于不可知的状态,不能保证能够从该服务的实例中获取反馈,所以需要服务剔除 AbstractInstanceRegistry#evict 方法定时清理这些不稳定的服务。该方法会批量将注册表中所有过期的租约剔除,实现代码如下所示。

```
    public void evict() {
        this.evict(0L);
    }

    public void evict(long additionalLeaseMs) {
        logger.debug("Running the evict task");
        if (!this.isLeaseExpirationEnabled()) {
            logger.debug("DS: lease expiration is currently disabled.");
        } else {
```

```java
            List<Lease<InstanceInfo>> expiredLeases = new ArrayList();
            Iterator var4 = this.registry.entrySet().iterator();

            while(true) {
                Map leaseMap;
                do {
                    if (!var4.hasNext()) {
                        int registrySize = (int)this.getLocalRegistrySize();
                        int registrySizeThreshold = (int)((double)registrySize * this.serverConfig.getRenewalPercentThreshold());
                        int evictionLimit = registrySize - registrySizeThreshold;
                        int toEvict = Math.min(expiredLeases.size(), evictionLimit);
                        if (toEvict > 0) {
                            logger.info("Evicting {} items (expired={}, evictionLimit={})", new Object[]{toEvict, expiredLeases.size(), evictionLimit});
                            Random random = new Random(System.currentTimeMillis());

                            for(int i = 0; i < toEvict; ++i) {
                                int next = i + random.nextInt(expiredLeases.size() - i);
                                Collections.swap(expiredLeases, i, next);
                                Lease<InstanceInfo> lease = (Lease)expiredLeases.get(i);
                                String appName = ((InstanceInfo)lease.getHolder()).getAppName();
                                String id = ((InstanceInfo)lease.getHolder()).getId();
                                EurekaMonitors.EXPIRED.increment();
                                logger.warn("DS: Registry: expired lease for {}/{}", appName, id);
                                this.internalCancel(appName, id, false);
                            }
                        }

                        return;
                    }

                    Entry<String, Map<String, Lease<InstanceInfo>>> groupEntry = (Entry)var4.next();
                    leaseMap = (Map)groupEntry.getValue();
                } while(leaseMap == null);

                Iterator var7 = leaseMap.entrySet().iterator();

                while(var7.hasNext()) {
                    Entry<String, Lease<InstanceInfo>> leaseEntry = (Entry)var7.next();
                    Lease<InstanceInfo> lease = (Lease)leaseEntry.getValue();
                    if (lease.isExpired(additionalLeaseMs) && lease.getHolder() != null) {
                        expiredLeases.add(lease);
                    }
                }
            }
        }
    }
```

服务剔除将会遍历注册表,找出其中所有的过期租约,然后根据配置文件中的续约百分比阈值和当前注册表的租约总数量计算出最大允许的剔除租约的数量(当前注册表中租约总数量减去当前注册表续约百分比阈值),分批剔除过期的服务实例租约。对过期的服务实例租约调用服务下线 **AbstractInstanceRegistry#**

internalCancel 方法将其从注册表中清除。

evict 方法中有很多限制，都是为了保证 Eureka 服务端的可用性，举例如下。

（1）自我保护时期不能进行服务剔除操作。

（2）过期操作是分批进行的。

（3）服务剔除是随机逐个剔除的，均匀分布在所有应用中，防止在同一时间、同一服务集群中的服务全部过期被剔除，以致大量剔除发生时，在未进行自我保护的情况下促使了程序的崩溃。

服务剔除是一个定时的任务，所以 AbstractInstanceRegistry 中定义了一个 EvictionTask 用于定时执行服务剔除，默认为 60s 一次。服务剔除的定时任务一般在 AbstractInstanceRegistry 初始化结束后进行，按照执行频率 EvictionIntervalTimerInMs 的设定，定时剔除过期的服务实例租约。

自我保护机制主要在 Eureka 客户端和 Eureka 服务端之间存在网络分区的情况下发挥保护作用，在服务端和客户端都有对应的实现。假设在某种特定的情况下，如网络故障，Eureka 客户端和 Eureka 服务端无法进行通信，此时 Eureka 客户端无法向 Eureka 服务端发起服务注册和续约请求，Eureka 服务端中注册表中的服务实例租约就可能因出现大量过期而面临被剔除的危险，然而此时的 Eureka 客户端可能是处于健康状态的（可接受服务访问），如果直接将注册表中大量过期的服务实例租约剔除显然是不合理的。

针对这种情况，Eureka 设计了自我保护机制。在 Eureka 服务端处，如果出现大量的服务实例过期被剔除的现象，那么该 Server 节点将进入自我保护模式，以保护注册表中的服务实例租约不再被剔除，在通信稳定后再退出该模式。在 Eureka 客户端处，如果向 Eureka 服务端注册失败，将快速超时并尝试与其他的 Eureka 服务端进行通信。自我保护机制的设计大大提高了 Eureka 的可用性。

4.5.5 服务下线

Eureka 客户端在应用销毁时，会向 Eureka 服务端发送服务下线请求，以清除注册表中关于本应用的租约，避免无效的服务被调用。在服务剔除的过程中，也是通过服务下线的方式完成对单个服务实例过期租约的剔除工作的。

服务下线的主要实现代码在 AbstractInstanceRegistry#internalCancel 方法中，该方法仅需要服务实例的服务名称和服务实例 id 即可完成服务下线。具体代码如下所示。

```
protected boolean internalCancel(String appName, String id, boolean isReplication) {
    boolean var7;
    try {
        this.read.lock();
        EurekaMonitors.CANCEL.increment(isReplication);
        Map<String, Lease<InstanceInfo>> gMap = (Map)this.registry.get(appName);
        Lease<InstanceInfo> leaseToCancel = null;
        if (gMap != null) {
            leaseToCancel = (Lease)gMap.remove(id);
        }

        synchronized(this.recentCanceledQueue) {
```

```
            this.recentCanceledQueue.add(new Pair(System.currentTimeMillis(), appName + "(" +
id + ")"));
        }

         InstanceStatus instanceStatus = (InstanceStatus)this.overriddenInstanceStatusMap.remove(id);
        if (instanceStatus != null) {
            logger.debug("Removed instance id {} from the overridden map which has value {}", id,
instanceStatus.name());
        }

        if (leaseToCancel != null) {
            leaseToCancel.cancel();
            InstanceInfo instanceInfo = (InstanceInfo)leaseToCancel.getHolder();
            String vip = null;
            String svip = null;
            if (instanceInfo != null) {
                instanceInfo.setActionType(ActionType.DELETED);
                this.recentlyChangedQueue.add(new AbstractInstanceRegistry.RecentlyChangedItem(leaseToCancel));
                instanceInfo.setLastUpdatedTimestamp();
                vip = instanceInfo.getVIPAddress();
                svip = instanceInfo.getSecureVipAddress();
            }

            this.invalidateCache(appName, vip, svip);
            logger.info("Cancelled instance {}/{} (replication={})", new Object[]{appName, id,
isReplication});
            boolean var10 = true;
            return var10;
        }

        EurekaMonitors.CANCEL_NOT_FOUND.increment(isReplication);
        logger.warn("DS: Registry: cancel failed because Lease is not registered for: {}/{}",
appName, id);
        var7 = false;
    } finally {
        this.read.unlock();
    }

    return var7;
}
```

internalCancel 方法的行为过程与 register 方法的很类似，首先在注册表中根据服务名称和服务实例 id 查询关于服务实例的租约 Lease 是否存在，统计最近请求下线的服务实例用于 Eureka 服务端主页展示。如果租约不存在，返回下线失败；如果租约存在，从注册表中移除，设置租约的下线时间，同时在最近租约变更记录队列中添加新的下线记录，以用于 Eureka 客户端增量式获取注册表信息，最后设置 Response 缓存过期。

internalCancel 方法中同样通过读锁保证注册表中信息的一致性，避免脏读。

4.5.6 集群同步

如果 Eureka 服务端通过集群的方式进行部署，那么为了维护整个集群中 Eureka 服务端注册表信息的一致性，势必需要一个机制来同步 Eureka 服务端集群中的注册表信息。Eureka 服务端集群同步包含两部分：一部分是 Eureka 服务端在启动过程中从它可能存在的 Peer 节点中拉取注册表信息，并将这些服务实例的信息注册到本地注册表中；另一部分是 Eureka 服务端每次对本地注册表进行操作时，将操作同步到它的 Peer 节点中，以达到集群注册表信息一致的目的。

■ Eureka 服务端初始化本地注册表信息

Eureka 服务端在启动的过程中，会从它可能存在的 Peer 节点中拉取注册表信息，并将其中的服务实例信息注册到本地注册表，这部分功能主要通过 PeerAwareInstanceRegistryImpl#syncUp 方法完成。代码如下所示。

```java
public int syncUp() {
    int count = 0;

    for(int i = 0; i < this.serverConfig.getRegistrySyncRetries() && count == 0; ++i) {
        if (i > 0) {
            try {
                Thread.sleep(this.serverConfig.getRegistrySyncRetryWaitMs());
            } catch (InterruptedException var10) {
                logger.warn("Interrupted during registry transfer");
                break;
            }
        }

        Applications apps = this.eurekaClient.getApplications();
        Iterator var4 = apps.getRegisteredApplications().iterator();

        while(var4.hasNext()) {
            Application app = (Application)var4.next();
            Iterator var6 = app.getInstances().iterator();

            while(var6.hasNext()) {
                InstanceInfo instance = (InstanceInfo)var6.next();

                try {
                    if (this.isRegisterable(instance)) {
                        this.register(instance, instance.getLeaseInfo().getDurationInSecs(), true);
                        ++count;
                    }
                } catch (Throwable var9) {
                    logger.error("During DS init copy", var9);
                }
            }
        }
    }

    return count;
}
```

Eureka 服务端也是 Eureka 客户端，在启动的时候也会进行 DiscoveryClient 的初始化，会从其对应的 Eureka 服务端中拉取全量的注册表信息。在 Eureka 服务端集群部署的情况下，Eureka 服务端从它的 Peer 节点中拉取到注册表信息后，将遍历这个 Applications，将所有的服务实例通过 Abstract Instance Registry#register 方法注册到本地注册表中。

在初始化本地注册表时，Eureka 服务端并不会接受来自 Eureka 客户端的通信请求（如注册、获取注册表信息等请求）。在同步注册表信息后会通过 PeerAwareInstanceRegistryImpl#openForTraffic 方法允许该 Server 接收流量。代码如下所示。

```java
public void openForTraffic(ApplicationInfoManager applicationInfoManager, int count) {
    this.expectedNumberOfClientsSendingRenews = count;
    this.updateRenewsPerMinThreshold();
    logger.info("Got {} instances from neighboring DS node", count);
    logger.info("Renew threshold is: {}", this.numberOfRenewsPerMinThreshold);
    this.startupTime = System.currentTimeMillis();
    if (count > 0) {
        this.peerInstancesTransferEmptyOnStartup = false;
    }

    Name selfName = applicationInfoManager.getInfo().getDataCenterInfo().getName();
    boolean isAws = Name.Amazon == selfName;
    if (isAws && this.serverConfig.shouldPrimeAwsReplicaConnections()) {
        logger.info("Priming AWS connections for all replicas");
        this.primeAwsReplicas(applicationInfoManager);
    }

    logger.info("Changing status to UP");
    applicationInfoManager.setInstanceStatus(InstanceStatus.UP);
    super.postInit();
}
```

在 Eureka 服务端中有一个过滤器 StatusFilter，用于检查 Eureka 服务端的状态，当其状态不为 UP 时，将拒绝所有的请求。在 Eureka 客户端请求获取注册表信息时，Eureka 服务端会判断此时是否允许获取注册表中的信息。上述做法是为了避免 Eureka 服务端在 syncUp 方法中没有获取任何服务实例信息时（Eureka 服务端集群部署的情况下），Eureka 服务端注册表中的信息影响到 Eureka 客户端缓存的注册表中的信息。如果 Eureka 服务端在 syncUp 方法中没有获得任何的服务实例信息，它将把 peerInstancesTransferEmptyOnStartup 设置为 true，这时该 Eureka 服务端在 WaitTimeInMsWhenSyncEmpty（可以通过 eureka.server.wait-time-in-ms-when-sync-empty 设置，默认为 5min）时间后才能被 Eureka 客户端访问以获取注册表信息。

▌ Eureka 服务端之间注册表信息的同步复制

为了保证 Eureka 服务端集群运行时注册表信息的一致性，每个 Eureka 服务端在对本地注册表进行操作时，会将相应的操作同步到所有 Peer 节点中。

在 PeerAwareInstanceRegistryImpl 中，对 AbstractInstanceRegistry 中的 register、renew 和 cancel 等方法都添加了同步到 Peer 节点的操作，使集群中的注册表信息保持最终一致性，代码如下所示。

```java
    public void register(InstanceInfo info, boolean isReplication) {
        int leaseDuration = 90;
        if (info.getLeaseInfo() != null && info.getLeaseInfo().getDurationInSecs() > 0) {
            leaseDuration = info.getLeaseInfo().getDurationInSecs();
        }

        super.register(info, leaseDuration, isReplication);
        this.replicateToPeers(PeerAwareInstanceRegistryImpl.Action.Register, info.getAppName(), info.getId(), info, (InstanceStatus)null, isReplication);
    }

    public boolean renew(String appName, String id, boolean isReplication) {
        if (super.renew(appName, id, isReplication)) {
            this.replicateToPeers(PeerAwareInstanceRegistryImpl.Action.Heartbeat, appName, id, (InstanceInfo)null, (InstanceStatus)null, isReplication);
            return true;
        } else {
            return false;
        }
    }
    public boolean cancel(String appName, String id, boolean isReplication) {
        if (super.cancel(appName, id, isReplication)) {
            this.replicateToPeers(PeerAwareInstanceRegistryImpl.Action.Cancel, appName, id, (InstanceInfo)null, (InstanceStatus)null, isReplication);
            synchronized(this.lock) {
                if (this.expectedNumberOfClientsSendingRenews > 0) {
                    --this.expectedNumberOfClientsSendingRenews;
                    this.updateRenewsPerMinThreshold();
                }
            }

            return true;
        } else {
            return false;
        }
    }
```

同步的操作使用枚举定义，如下所示。

```java
    public static enum Action {
        Heartbeat,
        Register,
        Cancel,
        StatusUpdate,
        DeleteStatusOverride;

        private com.netflix.servo.monitor.Timer timer = Monitors.newTimer(this.name());

        private Action() {
        }

        public com.netflix.servo.monitor.Timer getTimer() {
            return this.timer;
        }
    }
```

对此需要关注 replicateToPeers 方法，它将遍历 Eureka 服务端中的 Peer 节点，向每个 Peer 节点发送同步请求，代码如下所示。

```java
    private void replicateToPeers(PeerAwareInstanceRegistryImpl.Action action, String appName,
String id, InstanceInfo info, InstanceStatus newStatus, boolean isReplication) {
        Stopwatch tracer = action.getTimer().start();

        try {
            if (isReplication) {
                this.numberOfReplicationsLastMin.increment();
            }

            if (this.peerEurekaNodes != Collections.EMPTY_LIST && !isReplication) {
                Iterator var8 = this.peerEurekaNodes.getPeerEurekaNodes().iterator();

                while(var8.hasNext()) {
                    PeerEurekaNode node = (PeerEurekaNode)var8.next();
                    if (!this.peerEurekaNodes.isThisMyUrl(node.getServiceUrl())) {
                        this.replicateInstanceActionsToPeers(action, appName, id, info, newStatus, node);
                    }
                }

                return;
            }
        } finally {
            tracer.stop();
        }

    }
```

PeerEurekaNode 表示一个可同步共享数据的 Eureka 服务端。在 PeerEurekaNode 中，具有 Register、Cancel、Heartbeat 和 StatusUpdate 等诸多用于向 Peer 节点同步注册表信息的操作。

在 replicateInstanceActionsToPeers 方法中，根据 action 的不同，调用 PeerEurekaNode 的不同方法进行同步复制，代码如下所示。

```java
    private void replicateInstanceActionsToPeers(PeerAwareInstanceRegistryImpl.Action action,
String appName, String id, InstanceInfo info, InstanceStatus newStatus, PeerEurekaNode node) {
        try {
            InstanceInfo infoFromRegistry = null;
            CurrentRequestVersion.set(Version.V2);
            switch(action) {
            case Cancel:
                node.cancel(appName, id);
                break;
            case Heartbeat:
                InstanceStatus overriddenStatus = (InstanceStatus)this.overriddenInstanceStatusMap.get(id);
                infoFromRegistry = this.getInstanceByAppAndId(appName, id, false);
                node.heartbeat(appName, id, infoFromRegistry, overriddenStatus, false);
                break;
            case Register:
                node.register(info);
```

```
          break;
      case StatusUpdate:
          infoFromRegistry = this.getInstanceByAppAndId(appName, id, false);
          node.statusUpdate(appName, id, newStatus, infoFromRegistry);
          break;
      case DeleteStatusOverride:
          infoFromRegistry = this.getInstanceByAppAndId(appName, id, false);
          node.deleteStatusOverride(appName, id, infoFromRegistry);
      }
  } catch (Throwable var9) {
      logger.error("Can not replicate information to {} for action {}", new Object[]{node.getServiceUrl(), action.name(), var9});
  }
}
```

PeerEurekaNode 中的每一个同步复制操作都是通过批任务和流处理的方式完成的，同一时间段内相同服务实例的相同操作将使用相同的任务编号，在进行同步复制的时候根据任务编号合并操作，减少同步操作的数量和网络消耗，但是这样同时也造成同步复制的延时，不满足 CAP 中的 C（一致性）。

通过 Eureka 服务端在启动过程中初始化本地注册表信息和 Eureka 服务端集群间的同步复制操作，最终达到集群中 Eureka 服务端注册表信息一致的目的。

4.5.7 获取注册表中服务实例的信息

在 Eureka 服务端中获取注册表的服务实例信息主要通过两个方法来实现：AbstractInstanceRegistry#getApplicationsFromMultipleRegions 用于从多个地区获取全量注册表信息，AbstractInstanceRegistry#getApplicationDeltasFromMultipleRegions 用于从多个地区获取增量式注册表信息。

getApplicationsFromMultipleRegions 方法

getApplicationsFromMultipleRegions 方法会从多个地区中获取全量注册表信息，并将其封装成 Applications 返回。代码如下所示。

```
public Applications getApplicationsFromMultipleRegions(String[] remoteRegions) {
    boolean includeRemoteRegion = null != remoteRegions && remoteRegions.length != 0;
    logger.debug("Fetching applications registry with remote regions: {}, Regions argument {}", includeRemoteRegion, remoteRegions);
    if (includeRemoteRegion) {
        EurekaMonitors.GET_ALL_WITH_REMOTE_REGIONS_CACHE_MISS.increment();
    } else {
        EurekaMonitors.GET_ALL_CACHE_MISS.increment();
    }

    Applications apps = new Applications();
    apps.setVersion(1L);
    Iterator var4 = this.registry.entrySet().iterator();

    while(var4.hasNext()) {
        Entry<String, Map<String, Lease<InstanceInfo>>> entry = (Entry)var4.next();
```

```java
        Application app = null;
        Lease lease;
        if (entry.getValue() != null) {
                for(Iterator var7 = ((Map)entry.getValue()).entrySet().iterator(); var7.hasNext(); app.addInstance(this.decorateInstanceInfo(lease))) {
                Entry<String, Lease<InstanceInfo>> stringLeaseEntry = (Entry)var7.next();
                lease = (Lease)stringLeaseEntry.getValue();
                if (app == null) {
                   app = new Application(((InstanceInfo)lease.getHolder()).getAppName());
                }
            }
        }

        if (app != null) {
            apps.addApplication(app);
        }
    }

    if (includeRemoteRegion) {
        String[] var15 = remoteRegions;
        int var16 = remoteRegions.length;

        label69:
        for(int var17 = 0; var17 < var16; ++var17) {
            String remoteRegion = var15[var17];
                RemoteRegionRegistry remoteRegistry = (RemoteRegionRegistry)this.regionNameVSRemoteRegistry.get(remoteRegion);
            if (null == remoteRegistry) {
                logger.warn("No remote registry available for the remote region {}", remoteRegion);
            } else {
                Applications remoteApps = remoteRegistry.getApplications();
                Iterator var10 = remoteApps.getRegisteredApplications().iterator();

                while(true) {
                    while(true) {
                        if (!var10.hasNext()) {
                            continue label69;
                        }

                        Application application = (Application)var10.next();
                        if (this.shouldFetchFromRemoteRegistry(application.getName(), remoteRegion)) {
                                logger.info("Application {}  fetched from the remote region {}", application.getName(), remoteRegion);
                            Application appInstanceTillNow = apps.getRegisteredApplications(application.getName());

                            if (appInstanceTillNow == null) {
                               appInstanceTillNow = new Application(application.getName());
                               apps.addApplication(appInstanceTillNow);
                            }

                            Iterator var13 = application.getInstances().iterator();

                            while(var13.hasNext()) {
                               InstanceInfo instanceInfo = (InstanceInfo)var13.next();
                               appInstanceTillNow.addInstance(instanceInfo);
```

```
                    }
                } else {
                    logger.debug("Application {} not fetched from the remote region {} as there 
exists a whitelist and this app is not in the whitelist.", application.getName(), remoteRegion);
                }
            }
        }
    }
}

apps.setAppsHashCode(apps.getReconcileHashCode());
return apps;
}
```

上述代码首先将本地注册表中的所有服务实例信息提取出来封装到 Applications 中,再根据需要将远程区域的 Eureka 服务端注册表中的服务实例信息添加到 Applications 中。最后将封装了全量注册表信息的 Applications 返回给 Eureka 客户端。

■ getApplicationDeltasFromMultipleRegions 方法

getApplicationDeltasFromMultipleRegions 方法会从多个地区中获取增量式注册表信息,并将其封装成 Applications 返回。代码如下所示。

```java
public Applications getApplicationDeltasFromMultipleRegions(String[] remoteRegions) {
    if (null == remoteRegions) {
        remoteRegions = this.allKnownRemoteRegions;
    }

    boolean includeRemoteRegion = remoteRegions.length != 0;
    if (includeRemoteRegion) {
        EurekaMonitors.GET_ALL_WITH_REMOTE_REGIONS_CACHE_MISS_DELTA.increment();
    } else {
        EurekaMonitors.GET_ALL_CACHE_MISS_DELTA.increment();
    }

    Applications apps = new Applications();
    apps.setVersion(this.responseCache.getVersionDeltaWithRegions().get());
    HashMap applicationInstancesMap = new HashMap();

    Applications var23;
    try {
        this.write.lock();
        Iterator<AbstractInstanceRegistry.RecentlyChangedItem>iter=this.recentlyChangedQueue.iterator();
        logger.debug("The number of elements in the delta queue is:{}",this.recentlyChangedQueue.size());

        Lease lease;
        Application app;
        for(; iter.hasNext(); app.addInstance(new InstanceInfo(this.decorateInstanceInfo(lease)))) {
            lease = ((AbstractInstanceRegistry.RecentlyChangedItem)iter.next()).getLeaseInfo();
            InstanceInfo instanceInfo = (InstanceInfo)lease.getHolder();
```

```java
            logger.debug("The instance id {} is found with status {} and actiontype {}", new Object[]
{instanceInfo.getId(), instanceInfo.getStatus().name(), instanceInfo.getActionType().name()});
            app = (Application)applicationInstancesMap.get(instanceInfo.getAppName());
            if (app == null) {
                app = new Application(instanceInfo.getAppName());
                applicationInstancesMap.put(instanceInfo.getAppName(), app);
                apps.addApplication(app);
            }
        }
    }

    if (includeRemoteRegion) {
        String[] var20 = remoteRegions;
        int var22 = remoteRegions.length;

        label155:
        for(int var24 = 0; var24 < var22; ++var24) {
            String remoteRegion = var20[var24];
                RemoteRegionRegistry remoteRegistry = (RemoteRegionRegistry)this.
regionNameVSRemoteRegistry.get(remoteRegion);
            if (null != remoteRegistry) {
                Applications remoteAppsDelta = remoteRegistry.getApplicationDeltas();
                if (null != remoteAppsDelta) {
                    Iterator var12 = remoteAppsDelta.getRegisteredApplications().iterator();

                    while(true) {
                        Application application;
                        do {
                            if (!var12.hasNext()) {
                                continue label155;
                            }

                            application = (Application)var12.next();
                        } while(!this.shouldFetchFromRemoteRegistry(application.getName(),
remoteRegion));

                        Application appInstanceTillNow = apps.getRegisteredApplications(application.
getName());
                        if (appInstanceTillNow == null) {
                            appInstanceTillNow = new Application(application.getName());
                            apps.addApplication(appInstanceTillNow);
                        }

                        Iterator var15 = application.getInstances().iterator();

                        while(var15.hasNext()) {
                            InstanceInfo instanceInfo = (InstanceInfo)var15.next();
                            appInstanceTillNow.addInstance(new InstanceInfo(instanceInfo));
                        }
                    }
                }
            }
        }
    }

    Applications allApps = this.getApplicationsFromMultipleRegions(remoteRegions);
```

```
                apps.setAppsHashCode(allApps.getReconcileHashCode());
                var23 = apps;
        } finally {
            this.write.unlock();
        }

        return var23;
    }
```

获取增量式注册表信息将会从 recentlyChangedQueue 中获取最近变化的服务实例信息。recentlyChangedQueue 中统计了近 3min 内注册、修改和剔除的服务实例信息，在服务注册 AbstractInstanceRegistry#register、接收心跳请求 AbstractInstanceRegistry#renew 和服务下线 AbstractInstanceRegistry#internalCancel 等方法中均可见到 recentlyChangedQueue 对这些服务实例进行登记，用于记录增量式注册表信息。getApplicationsFromMultipleRegions 方法同样提供了从远程地区的 Eureka 服务端获取增量式注册表信息的能力。

第 5 章

基于 Ribbon 服务调用

目前已经实现了服务注册和服务发现,当启动某个服务的时候,可以通过 HTTP 的形式将服务注册到注册中心,并且可以通过 Spring Cloud 提供的工具获取注册中心的服务列表。但是服务之间的调用还存在很多的问题,比如如何更加方便地调用微服务,多个微服务的提供者如何选择。这些问题可以使用 Ribbon 来解决。

本章的主要内容如下。

1. 认识 Ribbon。
2. 基于 Ribbon 实现负载均衡调用。
3. Ribbon 源码解析。

5.1 认识 Ribbon

Spring Cloud Ribbon 是一个基于 HTTP 和 TCP 的客户端负载均衡工具,它基于 Netflix Ribbon 实现。通过 Spring Cloud 的封装,可以让使用者轻松地将面向服务的 REST 模板请求自动转换成客户端负载均衡的服务调用。

Ribbon 主要有以下三大子模块。

(1) Ribbon-Core: 该模块为 Ribbon 的核心,主要包括负载均衡器接口定义、客户端接口定义、内置的负载均衡实现等 API。

(2) Ribbon-Eureka: 该模块为 Eureka 客户端提供负载均衡实现类。

(3) Ribbon-HttpClient: 该模块提供含有负载均衡功能的 REST 客户端,用于对 Apache 的 HttpClient 进行封装。

5.1.1 微服务之间的交互

微服务提倡将一个原本独立的系统分成众多小型服务系统,这些小型服务系统都在独立的进程中运行,通过各个小型服务系统之间的协作来实现原本独立系统的所有业务功能。

小型服务系统使用多种跨进程的方式进行通信协作,而 RESTful 风格的网络请求是最为常见的交互方式之一。

RESTful 风格的网络请求中的 REST 是 Resource Representational State Transfer 的缩写,即资源表现层状态转移。

Resource 代表互联网资源。所谓资源是网络上的一个实体,或者说网上的一个具体信息。它可以是一段文本、一首歌曲、一种服务,可以使用一个 URL 指向它,每种资源对应一个 URL。

Representational 是表现层的意思。资源是一种消息实体,它可以有多种外在的表现形式,把资源具体呈现出来的形式叫作它的表现层。比如文本可以用 TXT 格式进行表现,也可以使用 XML 格式、JSON 格式和二进制格式;视频可以用 MP4 格式表现,也可以用 AVI 格式表现。URL 只代表资源的实体,不代表它的形式。它的具体表现形式应该由 HTTP 请求的头信息 Accept 和 Content-Type 字段指定,这两个字段是对表现层的描述。

State Transfer 是指状态转移。如果客户端想要操作服务端资源,必须通过某种手段,让服务端资源发生状态转移。客户端访问服务端的过程中必然涉及数据和状态的转化,而这种转化是建立在表现层之上的,所以被称为表现层状态转移。客户端通过使用 HTTP 中的 4 个方法来实现上述操作,它们分别是获取资源的 GET、新建资源的 POST、更新资源的 PUT 和删除资源的 DELETE。

5.1.2 Ribbon 的两个主要作用

Spring Cloud Ribbon 虽然只是一个工具类框架，不像注册中心、配置中心、API 网关那样需要独立部署，但是它几乎存在于每一个通过 Spring Cloud 构建的微服务和基础设施中。因为微服务间的调用、API 网关的请求转发等，实际上都是通过 Ribbon 来实现的，包括后续将要介绍的 Feign，也是基于 Ribbon 实现的工具。所以对 Spring Cloud Ribbon 的理解和使用，对于使用 Spring Cloud 来构建微服务非常重要。

Ribbon 的主要作用有以下两个。

（1）基于 Ribbon 实现服务调用，是通过拉取到的所有服务列表组成（服务名称请求路径的）映射关系，借助 RestTemplate 最终实现的。

（2）当有多个服务提供者时，Ribbon 可以根据负载均衡的算法自动选择需要调用的服务地址。

5.1.3 客户端的负载均衡

负载均衡在系统架构中是一个非常重要并且不得不去实现的内容。因为负载均衡是使系统高可用、缓解网络压力和扩容处理能力的重要手段之一。通常所说的负载均衡指的是服务端负载均衡，其中分为硬件负载均衡和软件负载均衡。硬件负载均衡主要通过在服务器节点之间安装专门用于负载均衡的设备来实现，比如 F5 等；而软件负载均衡则通过在服务器上安装一些具有负载均衡功能或模块的软件来实现，比如 Nginx 等。不论采用硬件负载均衡还是软件负载均衡，只要是服务端负载均衡都能以类似图 5-1 的方式构建起来。

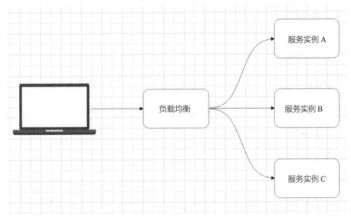

图5-1 负载均衡

硬件负载均衡的设备或软件负载均衡的软件模块都会维护一个可用的服务清单，通过心跳检测来剔除故障的服务节点以保证清单中都是可以正常访问的服务节点。当客户端发送请求到负载均衡设备的时候，该设备按某种算法（如线性轮询、按权重负载、按流量负载等）从其维护的可用服务清单中取出一台服务器的地址，然后进行转发。

客户端负载均衡和服务端负载均衡最大的不同点在于上面所提到的服务清单所存储的位置。在客户端负载均衡中，所有客户节点都维护着自己要访问的服务清单，这些服务清单来自注册中心。同服务端负载均衡的架构类似，在客户端负载均衡中也需要通过心跳去维护服务清单的"健康"，只是这个步骤需要与注册中心配合完成。在 Spring Cloud 实现的服务治理框架中，默认会创建针对各个服务治理框架的 Ribbon 自动化整合配置，比如 Eureka 中的 org.springframework.cloud.netflix.ribbon.eureka.RibbonEurekaAutoConfiguration，Consul 中的 org.springframework.cloud.consul.discovery.RibbonConsulAutoConfiguration。在实际使用的时候，可以通过查看这两个类的实现，找到它们的配置详情来更好地使用它们。

通过 Spring Cloud Ribbon 的封装，在微服务架构中使用客户端负载均衡非常简单，只需如下两步。

（1）服务提供者只需要启动多个服务实例并将其注册到一个注册中心或多个相关联的注册中心。

（2）服务消费者直接通过调用被 @LoadBalanced 注解修饰过的 RestTemplate 来实现面向服务的接口调用。

Ribbon 作为一个客户端负载均衡框架，默认的负载均衡策略是轮询，同时也提供了很多其他的策略让用户根据自身的业务需求进行选择。Ribbon 内部负责负载均衡的顶级接口为 com.netflix.loadbalancer.IRule。负载均衡策略接口实现类如图 5-2 所示。

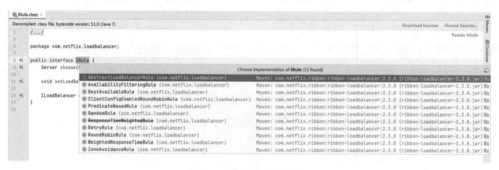

图5-2 负载均衡策略接口实现类

负载均衡策略接口实现类如下。

（1）BestAvailabl eRule：选择最小请求数的服务器，逐个考察服务器，如果服务器被标记为错误，则跳过，然后选择其中 ActiveRequestCount 最小的服务器。

（2）AvailabilityFilteringRule：过滤掉那些一直连接失败的且被标记为 circuit tripped 的后端服务器，并过滤掉那些高并发的后端服务器或者使用 AvailabilityPredicate 来实现过滤服务器的逻辑。其实就是检查 Status 里记录的各个服务器的运行状态。

（3）ZoneAvoidanceRule：分别使用 ZoneAvoidancePredicate 和 AvailabilityPredicate 来判断是否选择某个服务器，前一个判断一个服务区的运行状况是否可用，剔除不可用的服务区，后一个用于过滤连接数过多的 Server。

（4）RandomRule：随机选择一个服务器。

（5）RoundRobinRule：轮询选择，通过轮询 index，选择 index 对应位置的服务器。

（6）RetryRule：对选定的负载均衡策略机添加重试机制。也就是说，当选定了某个策略进行请求负载时，在一个配置时间段内若选择 Server 不成功，则一直尝试使用 subRule 的方式选择一个可用的 Server。

（7）ResponseTimeWeightedRule：作用同 WeightedResponseTimeRule。ResponseTimeWeightedRule 后来改名为 WeightedResponseTimeRule。

（8）WeightedResponseTimeRule：根据响应时间分配一个权重（Weight），响应时间越长，权重越小，被选中的可能性越低。

在服务消费者的 application.yml 配置文件中修改负载均衡策略，如下所示。

```
product-service:
  ribbon:
    NFLoadBalancerRuleClassName: com.netflix.loadbalancer.RandomRule
```

如果每个机器配置一样，则建议不修改负载均衡策略；如果部分机器配置强，则可以使用 WeightedResponseTimeRule。

5.2 基于 Ribbon 实现负载均衡调用

以 Eureka 为注册中心，完成订单微服务调用商品微服务。为了模拟负载均衡，启动两个商品微服务，端口号分别设置为 8001 和 8002。

5.2.1 坐标依赖

在 Spring Cloud 提供的服务发现的 JAR 包中已经包含了 Ribbon 的坐标依赖，如图 5-3 所示。org.springframework.cloud:spring-cloud-starter-netflix-eureka-client:2.1.0.RELEASE 包中已经包含了 org.springframework.cloud:spring-cloud-starter-netflix-ribbon:2.1.0.RELEASE，所以这里不需要导入任何额外的坐标。

图5-3　坐标依赖

5.2.2 工程改造

将贯穿案例的部分代码进行修改,这里需要注意的是 Ribbon 调用服务的时候,不支持带下画线的服务名称,所以需要将商品微服务名称中的下画线进行修改,建议将其修改为短横线。项目结构如图 5-4 所示。

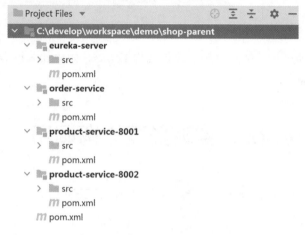

图5-4 项目结构

■ **修改端口号为 8001 的商品微服务对应的配置文件 application.yml**

修改端口号为 8001 的商品微服务对应的配置文件 application.yml,代码如下所示。

```yml
server:
  port: 8001 # 端口
spring:
  application:
    name: product-service # 服务名称
  datasource: # 数据源
    driver-class-name: com.mysql.jdbc.Driver
    url: jdbc:mysql://192.168.10.167:3306/shop?characterEncoding=utf8
    username: root
    password: root
  jpa:
    database: MySQL
    show-sql: true
    open-in-view: true
eureka:
  client:
    service-url:
      defaultZone: http://10.211.55.12:9000/eureka/
  instance:
    prefer-ip-address: true
    instance-id: ${spring.cloud.client.ip-address}:${server.port}
```

■ **修改端口号为 8002 的商品微服务对应的配置文件 application.yml**

修改端口号为 8002 的商品微服务对应的配置文件 application.yml,代码如下所示。

```yml
server:
```

```yaml
  port: 8002 # 端口
spring:
  application:
    name: product-service # 服务名称
  datasource: # 数据源
    driver-class-name: com.mysql.jdbc.Driver
    url: jdbc:mysql://192.168.10.167:3306/shop?characterEncoding=utf8
    username: root
    password: root
  jpa:
    database: MySQL
    show-sql: true
    open-in-view: true
eureka:
  client:
    service-url:
      defaultZone: http://10.211.55.12:9000/eureka/
  instance:
    prefer-ip-address:true
    instance-id: ${spring.cloud.client.ip-address}:${server.port}
```

■ 修改 order-service#OrderApplication 启动类

修改 order-service#OrderApplication 的启动类,代码如下所示。

```java
@SpringBootApplication
@EntityScan("cn.book.order.entity")
public class OrderApplication {
    @Bean
    @LoadBalanced
    public RestTemplate restTemplate(){
        return new RestTemplate();
    }

    public static void main(String[] args) {
        SpringApplication.run(OrderApplication.class, args);
    }
}
```

在 OrderApplication#restTemplate 方法之前加注解 @LoadBalanced,表示通过 restTemplate 方法根据负载均衡策略调用微服务。

■ 修改 product-service#ProductController#findById 方法

修改 product-service#ProductController#findById 方法,代码如下所示。

```java
@RestController
@RequestMapping("/product")
public class ProductController {

    @Autowired
    private ProductService productService;
```

```
@Value("${server.port}")
private String port;

//Spring Cloud 自动获取当前应用的 IP 地址
@Value("${spring.cloud.client.ip-address}")
private String ip;

@RequestMapping(value = "/{id}", method = RequestMethod.GET)
public Product findById(@PathVariable Long id) {
    Product product = productService.findById(id);
    product.setProductName(" 访问的服务地址信息: "+ip + ":" + port);
    return product;
}
```

为了可以直接地看到负载均衡的结果,在 product-service#ProductController#findById 方法中将微服务信息存入产品信息。

■ 修改 order-service#OrderController#order 方法

修改 order-service#OrderController#order 方法,代码如下所示。

```
@RestController
@RequestMapping("/order")
public class OrderController {

    @Autowired
    private OrderService orderService;

    /**
     * 注入 restTemplate 对象
     * 发送 HTTP 请求的工具类
     */
    @Autowired
    private RestTemplate restTemplate;

    @RequestMapping(value = "/{product_id}/{product_num}", method = RequestMethod.GET)
    public Map<String, Object> order(@PathVariable Long product_id, @PathVariable Integer product_num) {
        // 定义 URL
        String url = String.format("http://%s/product/%d", "product-service", product_id);
        // 使用 restTemplate 调用 HTTP 接口
        Product product = restTemplate.getForEntity(url, Product.class).getBody();
        // 创建字典对象,存放信息
        Map<String, Object> map = new HashMap<String, Object>();
        // 存数据
        map.put("product_num", product_num);
        map.put("product", product);
        // 返回数据
        return map;
    }
}
```

上述代码中的 URL 直接使用微服务的名称,不需要再使用 DiscoveryClient。端口号为 8001 和 8002 的商品微服务名称均为 product-service,Spring Cloud Ribbon 会自动根据微服务的名称 product-service 找到对应的微服务，并完成接口的调用。

5.2.3 代码测试

重启各个服务后，访问注册中心管理后台，结果如图 5-5 所示。

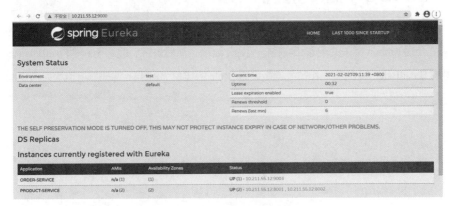

图5-5 注册中心管理后台

微服务 PRODUCT-SERVICE 对应的 Status 显示了模拟的两个信息。

访问订单微服务，结果如图 5-6 所示。

```
← → C  ▲ 不安全 | 10.211.55.12:9003/order/1/2
{
  - product: {
      id: 1,
      productName: "访问的服务地址信息：10.211.55.12:8001",
      status: 1,
      price: 6666,
      productDesc: "华为mate40",
      caption: "华为mate40",
      inventory: 100
    },
    product_num: 2
}
```

图5-6 访问订单微服务

再次访问订单微服务，结果如图 5-7 所示。

```
← → C  ▲ 不安全 | 10.211.55.12:9003/order/1/2
{
  - product: {
      id: 1,
      productName: "访问的服务地址信息：10.211.55.12:8002",
      status: 1,
      price: 6666,
      productDesc: "华为mate40",
      caption: "华为mate40",
      inventory: 100
    },
    product_num: 2
}
```

图5-7 再次访问订单微服务

通过两次访问订单微服务可以发现，订单微服务以轮询的方式调用了商品微服务。Ribbon 服务调用和负载均衡测试完成。

5.3 Ribbon 源码解析

本节首先介绍 Ribbon 相关的配置和实例的初始化过程，然后讲解负载均衡器（LoadBalancerClient），最后依次讲解 ILoadBalancer 的实现和负载均衡策略 IRule 的实现。Ribbon 源码解析流程如图 5-8 所示。

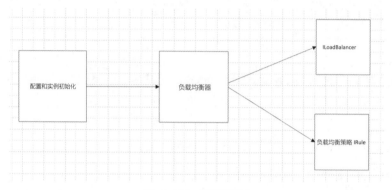

图5-8 Ribbon源码解析流程

5.3.1 配置和实例初始化

@RibbonClient 注解可以声明 Ribbon 客户端，配置 Ribbon 客户端的名称和配置类，Configuration 属性可以指定 @Configuration 的配置类，进行 Ribbon 相关的配置。@RibbonClient 还会导入 RibbonClientConfigurationRegistrar 类来动态注册 Ribbon 相关的 BeanDefinition。@RibbonClient 注解的具体实现如下所示。

```
@Import({RibbonClientConfigurationRegistrar.class})
public @interface RibbonClient {
    String value() default "";

    /**
     * 配置 Ribbon 客户端名称
     */
    String name() default "";

    /**
     * Ribbon 客户端的自定义配置，可以配置生成客户端的各个组件，如 ILoadBalancer、ServerListFilter
     * 和 IRule。默认的配置为 RibbonClientConfiguration.java
     */
    Class<?>[] configuration() default {};
}
```

RibbonClientConfigurationRegistrar 是 ImportBeanDefinitionRegistrar 的实现类，ImportBeanDefinitionRegistrar 是 Spring 动态注册 BeanDefinition 的接口，可以用来注册 Ribbon 所需的 BeanDefinition，如 Ribbon 客户端实

例（RibbonClient）。ImportBeanDefinitionRegistrar 的 registerBeanDefinitions 方法可以注册 Ribbon 客户端的配置类，也就是@ RibbonClient 的 Configuration 属性值。

registerBeanDefinitions 方法的具体实现如下所示。

```
//RibbonClientConfigurationRegistrar
public void registerBeanDefinitions(AnnotationMetadata metadata, BeanDefinitionRegistry registry) {
......
// 获取 @RibbonClient 的参数，获取 clientName 后进行 configuration 的注册
    Map<String, Object> client = metadata.getAnnotationAttributes(RibbonClient.class.getName(), true);
// 获取 RibbonClient 的 value 或者 name
    String name = this.getClientName(client);
    if (name != null) {
      this.registerClientConfiguration(registry, name, client.get("configuration"));
    }

}
```

Ribbon 对组件实例的管理配置机制和 OpenFeign 相同，都是通过 NamedContextFactory 创建带名称的 AnnotationConfigApplicationContext 来存储并管理不同的组件实例。

registerClientConfiguration 方法会向 BeanDefinitionRegistry 注册一个 RibbonClientSpecification 的 BeanDefinition，其名称为 RibbonClient 的名称加上 RibbonClientSpecification。RibbonClientSpecification 是 NamedContextFactory.Specification 的实现类，是供 SpringClientFactory 使用的。在 RibbonAutoConfiguration 里会进行 SpringClientFactory 实例的初始化，并设置将所有的 RibbonClientSpecification 实例都给 SpringClientFactory，供其在初始化 Ribbon 相关组件实例时使用，如下所示。

```
//RibbonClientConfigurationRegistrar.java
private void registerClientConfiguration(BeanDefinitionRegistry registry, Object name, Object configuration) {
    BeanDefinitionBuilder builder = BeanDefinitionBuilder.genericBeanDefinition(RibbonClientSpecification.class);
    builder.addConstructorArgValue(name);
    builder.addConstructorArgValue(configuration);
    registry.registerBeanDefinition(name + ".RibbonClientSpecification", builder.getBeanDefinition());
}
```

RibbonAutoConfiguration 配置类也会进行 LoadBalancerClient 接口的默认实例的初始化。若 loadBalancerClient 方法被@ ConditionalOnMissingBean 注解修饰，则意味着只有当 Spring 容器中没有 LoadBalancerClient 实例时，该方法才会初始化 RibbonLoadBalancerClient 对象，将其作为 LoadBalancerClient 的实例，如下所示。

```
@Configuration
@Conditional({RibbonAutoConfiguration.RibbonClassesConditions.class})
@RibbonClients
@AutoConfigureAfter(
    name = {"org.springframework.cloud.netflix.eureka.EurekaClientAutoConfiguration"}
)
```

```
//@AutoConfigureBefore 表明该配置会在 LoadBalancerAutoConfiguration 配置类之前
执行，因为后者会依赖前者
@AutoConfigureBefore({LoadBalancerAutoConfiguration.class, AsyncLoadBalancerAutoConfigurati
on.class})
@EnableConfigurationProperties({RibbonEagerLoadProperties.class, ServerIntrospectorProperties.
class})
public class RibbonAutoConfiguration {
// 在 RibbonClientConfigurationRegistrar 中注册的 RibbonClientSpecification 实例都会被注入这里
    @Autowired(
        required = false
    )
    private List<RibbonClientSpecification> configurations = new ArrayList();
    @Autowired
    private RibbonEagerLoadProperties ribbonEagerLoadProperties;

    public RibbonAutoConfiguration() {
    }

    @Bean
    public HasFeatures ribbonFeature() {
        return HasFeatures.namedFeature("Ribbon", Ribbon.class);
    }
//LoadBalancerClient 是核心的类
    @Bean
    @ConditionalOnMissingBean({LoadBalancerClient.class})
    public LoadBalancerClient loadBalancerClient() {
        return new RibbonLoadBalancerClient(this.springClientFactory());
    }
    ......
}
```

5.3.2 负载均衡器

本小节主要讲解负载均衡器——LoadBalancerClient 进行负载均衡的具体原理和实现。LoadBalancerClient 是 Ribbon 的核心类之一，可以在 RestTemplate 发送网络请求时替代 RestTemplate 进行网络调用。LoadBalancerClient 的定义如下所示。

```
//LoadBalancerClient.java
public interface LoadBalancerClient extends ServiceInstanceChooser {
// 从 serviceId 所代表的服务列表中选择一台服务器来发送网络请求
<T> T execute(String serviceId, LoadBalancerRequest<T> request) throws IOException;
<T> T execute(String serviceId, ServiceInstance serviceInstance, LoadBalancerRequest<T>
request) throws IOException;
// 构建网络请求 URI
URI reconstructURI(ServiceInstance instance, URI original);
}
```

LoadBalancerClient 接口继承了 ServiceInstanceChooser 接口，可以从服务器列表中根据负载均衡策略选出一个服务器实例。ServiceInstanceChooser 的定义如下所示。

```
// 实现该类来选择一台服务器用于发送请求
public interface ServiceInstanceChooser {
```

```
// 根据 serviceId 从服务器列表中选择一个 ServiceInstance
ServiceInstance choose(String serviceId);
}
```

RibbonLoadBalancerClient 是 LoadBalancerClient 的实现类之一，它的 execute 方法会首先使用 ILoadBalancer 来选择服务器实例（server），然后将该服务器实例封装成 RibbonServer 对象，最后调用 LoadBalancerRequest 的 apply 方法进行网络请求的处理。execute 方法的具体实现如下所示。

```
// RibbonLoadBalancerClient.java
public <T> T execute(String serviceId, LoadBalancerRequest<T> request, Object hint) throws IOException {
// 每次发送请求都会获取一个 ILoadBalancer 涉及负载均衡策略（IRule）、服务器列表（ServerList）
和检验服务是否存在（IPing）等细节实现
ILoadBalancer loadBalancer = getLoadBalancer(serviceId);
Server server = getServer(loadBalancer, hint);
if (server == null) {
throw new IllegalStateException("No instances available for " + serviceId);
}
RibbonServer ribbonServer = new RibbonServer(serviceId, server, isSecure(server,serviceId),
serverIntrospector(serviceId).getMetadata(server));

return execute(serviceId, ribbonServer, request);
}
```

getLoadBalancer 方法直接调用了 SpringClientFactory 的 getLoadBalancer 方法。SpringClientFactory 是 NamedContextFactory 的实现类。getLoadBalancer 方法的代码如下所示。

```
// RibbonLoadBalancerClient.java
protected ILoadBalancer getLoadBalancer(String serviceId) {
return this.clientFactory.getLoadBalancer(serviceId);
}
```

getServer 方法则直接调用了 ILoadBalancer 的 chooseServer 方法来使用负载均衡策略——从已知的服务器列表中选出一个服务器实例。gerServer 方法的具体实现如下所示。

```
// RibbonLoadBalancerClient.java
protected Server getServer(ILoadBalancer loadBalancer, Object hint) {
if (loadBalancer == null) {
return null;
}
// 对空指针使用 'default' 或只是传递指针？
return loadBalancer.chooseServer(hint != null ? hint : "default");
}
```

execute 方法调用 LoadBalancerRequest 实例的 apply 方法，将之前根据负载均衡策略选择出来的服务器作为参数传递进去，进行真正的 HTTP 请求发送，代码如下所示。

```
// RibbonLoadBalancerClient.java
public <T> T execute(String serviceId, ServiceInstance serviceInstance, LoadBalancerRequest<T>
request) throws IOException {
    Server server = null;
```

```
    if (serviceInstance instanceof RibbonLoadBalancerClient.RibbonServer) {
        server = ((RibbonLoadBalancerClient.RibbonServer)serviceInstance).getServer();
    }

    if (server == null) {
        throw new IllegalStateException("No instances available for " + serviceId);
    } else {
        RibbonLoadBalancerContext context = this.clientFactory.getLoadBalancerContext(serviceId);
        RibbonStatsRecorder statsRecorder = new RibbonStatsRecorder(context, server);

        try {
            T returnVal = request.apply(serviceInstance);
            statsRecorder.recordStats(returnVal);
            return returnVal;
        } catch (IOException var8) {
            statsRecorder.recordStats(var8);
            throw var8;
        } catch (Exception var9) {
            statsRecorder.recordStats(var9);
            ReflectionUtils.rethrowRuntimeException(var9);
            return null;
        }
    }
}
```

5.3.3 ILoadBalancer 的实现

ILoadBalancer 是 Ribbon 的关键类之一，是定义负载均衡操作过程的接口。Ribbon 通过 SpringClientFactory 类的 getLoadBalancer 方法可以获取 ILoadBalancer 实例。根据 Ribbon 的组件实例化机制，ILoadBalnacer 实例是在 RibbonAutoConfiguration 中被创建的。

SpringClientFactory 中的实例都是 RibbonClientConfiguration 或者自定义 Configuration 配置类创建的 Bean 实例。RibbonClientConfiguration 还创建了 IRule、IPing 和 ServerList 等相关组件的实例。使用者可以通过自定义配置类给出上述几个组件的不同实例。

如图 5-9 所示，ZoneAwareLoadBalancer 是 ILoadBalancer 接口的实现类之一，它是 Ribbon 默认的 ILoadBalancer 接口的实例。

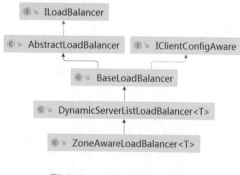

图5-9 ZoneAwareLoadBalancer类图

RibbonClientConfiguration 中有关 ZoneAwareLoadBalancer 的配置如下所示。

```java
// RibbonClientConfiguration.java
@Bean
@ConditionalOnMissingBean
public ILoadBalancer ribbonLoadBalancer(IClientConfig config, ServerList<Server> serverList,
 ServerListFilter<Server> serverListFilter, IRule rule, IPing ping, ServerListUpdater serverListUpdater)
{
    return (ILoadBalancer)(this.propertiesFactory.isSet(ILoadBalancer.class, this.name) ?
(ILoadBalancer)this.propertiesFactory.get(ILoadBalancer.class, config, this.name) : new
 ZoneAwareLoadBalancer(config, rule, ping, serverList, serverListFilter, serverListUpdater));
}
```

图 5-10 所示是与 IBalancer 相关的类图，其中的类都是 ZoneAwareLoadBalancer 构造方法所需参数实例的类。

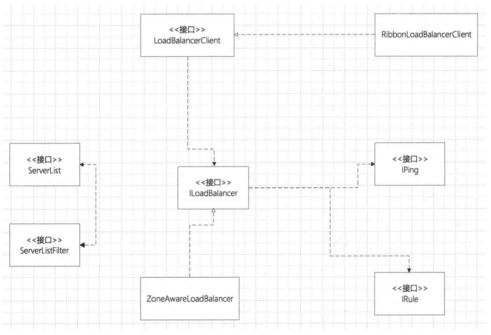

图5-10　与IBalancer相关的类图

接下来，按照 ZoneAwareLoadBalancer 构造方法的参数顺序来看一下与 ILoadBalancer 相关的重要的 Bean，它们分别是 IClientConfig、IRule、IPing、ServerList 和 ServerListFilter。默认 Bean 信息如表 5-1 所示。

表 5-1　默认 Bean 信息

Bean 类型	Bean 名称	类名	解释
IClientConfig	ribbonClientConfig	ribbonClientConfig	Client 的配置类
IRule	ribbonRule	RoundRobinRule	负载均衡策略
IPing	ribbonPing	DummyPing	服务可用性检测
ServerList	ribbonServerList	ConfigurationBasedServerList	服务器列表的获取
ServerListFilter	ribbonServerListFilter	ZonePreferenceServerListFilter	服务器列表的过滤

ZoneAwareLoadBalancer 的 chooseServer 方法首先使用 DynamicPropertyFactory 来获取平均负载（triggeringLoadPerServerThreshold）和平均实例故障率（AvoidZoneWithBlackoutPercentage）两个阈值，然后调用 ZoneAvoidanceRule 的 getAvailableZones 方法使用这两个阈值来获取所有可用的服务区（Zone）列表，每个服务区中包含了一定数量的服务器实例，接着调用 ZoneAvoidanceRule 的 randomChooseZone 方法从上述的服务区列表中随机选出一个服务区，最后调用该服务区对应 BaseLoadBalancer 实例的 chooseServer 方法获取最终的服务器实例。ZoneAwareLoadBalancer 会为不同的服务区调用不同的 BaseLoadBalancer 的 chooseServer 方法，这正体现了它类名的含义。chooseServer 方法是 BaseLoadBalancer 中最重要的方法之一，具体实现如下所示。

```java
//ZoneAwareLoadBalancer.java
public Server chooseServer(Object key) {
    if (ENABLED.get() && this.getLoadBalancerStats().getAvailableZones().size() > 1) {
        Server server = null;

        try {

// 获取当前有关负载均衡的服务器状态集合

            LoadBalancerStats lbStats = this.getLoadBalancerStats();
            Map<String, ZoneSnapshot> zoneSnapshot = ZoneAvoidanceRule.createSnapshot(lbStats);
            logger.debug("Zone snapshots: {}", zoneSnapshot);
// 使用 DynamicPropertyFactory 获取平均负载的阈值
            if (this.triggeringLoad == null) {
                this.triggeringLoad = DynamicPropertyFactory.getInstance().getDoubleProperty("ZoneAwareNIWSDiscoveryLoadBalancer." + this.getName() + ".triggeringLoadPerServerThreshold", 0.2D);
            }

// 使用 DynamicPropertyFactory 获取平均实例故障率的阈值

            if (this.triggeringBlackoutPercentage == null) {
                this.triggeringBlackoutPercentage = DynamicPropertyFactory.getInstance().getDoubleProperty("ZoneAwareNIWSDiscoveryLoadBalancer." + this.getName() + ".avoidZoneWithBlackoutPercentage", 0.99999D);
            }

// 根据两个阈值来获取所有可用的服务区列表

            Set<String> availableZones = ZoneAvoidanceRule.getAvailableZones(zoneSnapshot, this.triggeringLoad.get(), this.triggeringBlackoutPercentage.get());
            logger.debug("Available zones: {}", availableZones);
            if (availableZones != null && availableZones.size() < zoneSnapshot.keySet().size()) {

// 随机从可用的服务区列表中选择一个服务区

                String zone = ZoneAvoidanceRule.randomChooseZone(zoneSnapshot, availableZones);
                logger.debug("Zone chosen: {}", zone);
                if (zone != null) {
                    BaseLoadBalancer zoneLoadBalancer = this.getLoadBalancer(zone);
                    server = zoneLoadBalancer.chooseServer(key);
```

```
            }
          }
        } catch (Exception var8) {
          logger.error("Error choosing server using zone aware logic for load balancer={}", this.name, var8);
        }

        if (server != null) {
          return server;
        } else {
          logger.debug("Zone avoidance logic is not invoked.");
          return super.chooseServer(key);
        }
      } else {
        logger.debug("Zone aware logic disabled or there is only one zone");
        return super.chooseServer(key);
      }
    }
```

BaseLoadBalancer 对象的 chooseServer 方法实现比较简单，就是直接调用它的 IRule 成员变量的 choose 方法。IRule 是负责实现负载均衡策略的接口。BaseLoadBalancer 的 chooseServer 方法的代码如下所示。

```
// BaseLoadBalancer.java
public Server chooseServer(Object key) {
    if (this.counter == null) {
        this.counter = this.createCounter();
    }

    this.counter.increment();
    if (this.rule == null) {
        return null;
    } else {
        try {
            return this.rule.choose(key);
        } catch (Exception var3) {
            logger.warn("LoadBalancer [{}]: Error choosing server for key {}", new Object[]{this.name, key, var3});
            return null;
        }
    }
}
```

5.3.4 负载均衡策略实现

IRule 是定义 Ribbon 负载均衡策略的接口，可以通过实现该接口来自定义负载均衡策略，RibbonClientConfiguration 配置类则会给出 IRule 的默认实例。IRule 接口的 choose 方法可以从一堆服务器中根据一定规则选出一个服务器。IRule 有很多默认的实现类，这些实现类根据不同的算法和逻辑进行负载均衡。

图 5-11 展示了 Ribbon 提供的 IRule 的类图。

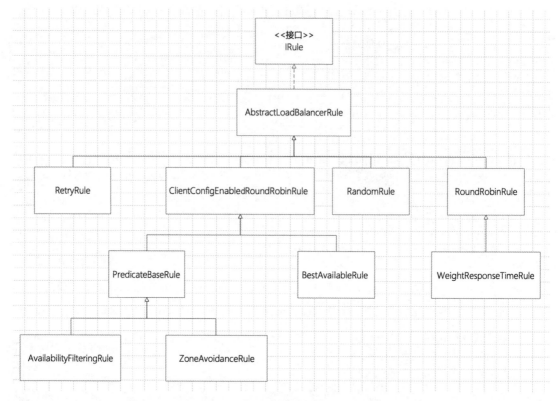

图5-11 IRule的类图

在大多数情况下，这些默认的实现类是可以满足需求的，如果有特殊需求，可以自己实现。Ribbon 内置的 IRule 子类如下所示。

（1）BestAvailableRule：选择最小请求数的服务器。

（2）ClientConfigEnabledRoundRobinRule：使用 RoundRobinRule 随机选择一个服务器。

（3）RoundRobinRule：以 RandomRobin 方法轮询选择服务器。

（4）RetryRule：在选定的负载均衡策略上添加重试机制。

（5）WeightedResponseTimeRule：根据响应时间计算一个权重，响应时间越长，权重越低，权重越低的服务器，被选择的可能性就越低。

（6）ZoneAvoidanceRule：根据服务器所属服务区的整体运行状况来轮询选择服务器。

Ribbon 的负载均衡策略既有 RoundRobinRule 和 RandomRule 这样不依赖于服务器运行状况的策略，也有 AvailabilityFilteringRule 和 WeightedResponseTimeRule 等多种基于服务器运行状况进行决策的策略。这些策略既可以依据单个服务器的运行状况，也可以依据整个服务区的运行状况来选择具体调用的服务器，几乎能满足各种场景需求。

ZoneAvoidanceRule 是 Ribbon 默认的 IRule 实例，较为复杂，在本小节的后续部分进行讲解。而 ClientConfigEnabledRoundRobinRule 是比较常用的 IRule 的子类之一，它使用的负载均衡策略是最为常见的 RoundRobin Rule，即简单轮询策略。其具体实现如下所示。

```java
public class ClientConfigEnabledRoundRobinRule extends AbstractLoadBalancerRule {
    RoundRobinRule roundRobinRule = new RoundRobinRule();

    public ClientConfigEnabledRoundRobinRule() {
    }

    public void initWithNiwsConfig(IClientConfig clientConfig) {
        this.roundRobinRule = new RoundRobinRule();
    }

    public void setLoadBalancer(ILoadBalancer lb) {
        super.setLoadBalancer(lb);
        this.roundRobinRule.setLoadBalancer(lb);
    }

    public Server choose(Object key) {
        if (this.roundRobinRule != null) {
            return this.roundRobinRule.choose(key);
        } else {
            throw new IllegalArgumentException("This class has not been initialized with the RoundRobinRule class");
        }
    }
}
```

RoundRobinRule 以轮询的方式依次选择不同的服务器,从序号为 1 的服务器开始,直到序号为 N 的服务器;下次选择服务器时,依然从序号为 1 的服务器开始。但是 RoundRobinRule 选择服务器时没有考虑服务器的状态,而 ZoneAvoidanceRule 选择时则会考虑服务器的状态,从而更好地进行负载均衡。

ZoneAvoidanceRule 是 Ribbon 默认的 IRule 实例,根据服务区的运行状况和服务器的可用性选择服务器,如下所示。

```java
// RibbonClientConfiguration.java
@Bean
@ConditionalOnMissingBean
public IRule ribbonRule(IClientConfig config) {

    // 如果在配置中设置了 Rule,则返回该 Rule,否则使用默认的 ZoneAvoidanceRule

    if (this.propertiesFactory.isSet(IRule.class, name)) {
        return this.propertiesFactory.get(IRule.class, config, name);
    }
    ZoneAvoidanceRule rule = new ZoneAvoidanceRule();
    rule.initWithNiwsConfig(config);
    return rule;
}
```

ZoneAvoidanceRule 的类图如图 5-12 所示。

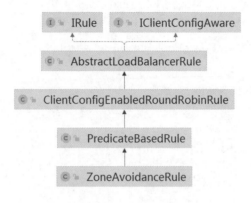

图5-12 ZoneAvoidanceRule的类图

ZoneAvoidanceRule 是根据服务器所属服务区的运行状况和服务器的可用性来进行负载均衡的。PredicateBasedRule 是 ZoneAvoidanceRule 的基类，它选择服务器的策略是先使用 ILoadBalancer 获取服务器列表，再使用 AbstractServerPredicate 进行服务器过滤，最后使用轮询策略从剩余的服务器列表中选出最终的服务器。PredicateBasedRule 的具体实现如下所示。

```java
public abstract class PredicateBasedRule extends ClientConfigEnabledRoundRobinRule {
    public PredicateBasedRule() {
    }

    public abstract AbstractServerPredicate getPredicate();

    public Server choose(Object key) {
        ILoadBalancer lb = this.getLoadBalancer();
        Optional<Server> server = this.getPredicate().chooseRoundRobinAfterFiltering(lb.getAllServers(), key);
        return server.isPresent() ? (Server)server.get() : null;
    }
}
```

PredicateBasedRule 的 getPredicate 方法需要子类来实现，不同的子类提供不同的 AbstractServerPredicate 实例来实现不同的服务器过滤策略。而 ZoneAvoidanceRule 的 getPredicate 方法的相关实现如下所示。

```java
//ZoneAvoidanceRule.class
public ZoneAvoidanceRule() {
    ZoneAvoidancePredicate zonePredicate = new ZoneAvoidancePredicate(this);
    AvailabilityPredicate availabilityPredicate = new AvailabilityPredicate(this);
    this.compositePredicate = this.createCompositePredicate(zonePredicate, availabilityPredicate);
}

// 将两个策略组合成 CompositePredicate
private CompositePredicate createCompositePredicate(ZoneAvoidancePredicate p1, AvailabilityPredicate p2) {
    return CompositePredicate.withPredicates(new AbstractServerPredicate[]{p1, p2}).addFallbackPredicate(p2).addFallbackPredicate(AbstractServerPredicate.alwaysTrue()).build();
}
```

```java
// 返回构造方法中生成的 compositePredicate

public AbstractServerPredicate getPredicate() {
    return this.compositePredicate;
}
```

CompositePredicate 的 chooseRoundRobinAfterFiltering 方法继承父类 AbstractServerPredicate 的实现。它首先调用 getEligibleServers 方法过滤服务器列表，然后使用轮询策略选择一个服务器返回，如下所示。

```java
//AbstractServerPredicate.java
public Optional<Server> chooseRoundRobinAfterFiltering(List<Server> servers) {
    List<Server> eligible = getEligibleServers(servers);
    if (eligible.size() == 0) {
        return Optional.absent();
    }
    return Optional.of(eligible.get(incrementAndGetModulo(eligible.size())));
}
```

当 loadBalancerKey 参数为 null 时，getEligibleServers 方法会使用 serverOnlyPredicate 来依次过滤服务器列表。getEligibleServers 方法的具体实现如下所示。

```java
public List<Server> getEligibleServers(List<Server> servers, Object loadBalancerKey) {
    if (loadBalancerKey == null) {
        return ImmutableList.copyOf(Iterables.filter(servers, this.getServerOnlyPredicate()));
    } else {
        List<Server> results = Lists.newArrayList();
        Iterator var4 = servers.iterator();

        // 遍历 servers，调用对应策略的 apply 方法判断该服务器是否可用

        while(var4.hasNext()) {
            Server server = (Server)var4.next();
            if (this.apply(new PredicateKey(loadBalancerKey, server))) {
                results.add(server);
            }
        }

        return results;
    }
}
```

serverOnlyPredicate 会调用其 apply 方法，并将 Server 对象封装成 PredicateKey 作为参数传入。AbstractServerPredicate 并没有实现 apply 方法，而是由它的子类来实现，从而达到不同子类实现不同过滤策略的目的。serverOnlyPredicate 的代码如下所示。

```java
//AbstractServerPredicate.java
private final Predicate<Server> serverOnlyPredicate =  new Predicate<Server>() {
    @Override
    public boolean apply(@Nullable Server input) {
        return AbstractServerPredicate.this.apply(new PredicateKey(input));
    }
};
```

ZoneAvoidanceRule 类中 CompositePredicate 对象的 apply 方法会依次调用 ZoneAvoidancePredicate 和 AvailabilityPredicate 的 apply 方法。

ZoneAvoidancePredicate 以服务区为单位考察所有服务区的整体运行状况，对于不可用的服务区整个丢弃，从剩下的服务区中选可用的服务器，并且会判断出最差的服务区并排除。ZoneAvoidancePredicate 的 apply 方法代码如下所示。

```java
public boolean apply(@Nullable PredicateKey input) {
    if (!ENABLED.get()) {
        return true;
    } else {
        String serverZone = input.getServer().getZone();
        if (serverZone == null) {
// 如果服务器没有服务区相关的信息，则直接返回
            return true;
        } else {
//LoadBalancerStats 存储着每个服务器或者节点的执行特征和运行记录
// 这些信息可供负载均衡策略使用
            LoadBalancerStats lbStats = this.getLBStats();
            if (lbStats == null) {
                return true;
            } else if (lbStats.getAvailableZones().size() <= 1) {
// 只有一个服务区的时候也直接返回
                return true;
            } else {
// 为了效率，先看一下 lbStats 中记录的服务区列表是否包含当前服务区
                Map<String, ZoneSnapshot> zoneSnapshot = ZoneAvoidanceRule.createSnapshot(lbStats);
                if (!zoneSnapshot.keySet().contains(serverZone)) {
// 如果当前服务区不存在，那么也直接返回
                    return true;
                } else {
                    logger.debug("Zone snapshots: {}", zoneSnapshot);
// 调用 ZoneAvoidanceRule 的 getAvailableZones 方法来获取可用的服务区列表
                    Set<String> availableZones = ZoneAvoidanceRule.getAvailableZones(zoneSnapshot, this.triggeringLoad.get(), this.triggeringBlackoutPercentage.get());
                    logger.debug("Available zones: {}", availableZones);
// 判断当前服务区是否在可用服务区列表中
```

```
                return availableZones != null ? availableZones.contains(input.getServer().getZone()) :
false;
            }
        }
    }
}
```

服务区是多个服务实例的集合，不同服务区之间的服务实例相互访问会有更大的网络延迟，服务区内的服务实例相互访问网络延迟较小。ZoneSnapshot 存储了关于服务区的部分运行状况数据，比如说实例数、平均负载、断路器断开数、活动请求数等。ZoneSnapshot 的定义如下所示。

```
//ZoneSnapshot.java
public class ZoneSnapshot {
    final int instanceCount; // 实例数
    final double loadPerServer; // 平均负载
    final int circuitTrippedCount; // 断路器断开数
    final int activeRequestsCount; // 活动请求数
    ......
}
```

ZoneAvoidanceRule 的 createSnapshot 方法其实就是将所有的服务区列表转化为以其名称为键值的哈希表，供 ZoneAvoidancePredicate 的 apply 方法使用，如下所示。

```
// ZoneAvoidanceRule.java
// 将 LoadBalancerStats 中的 availableZones 列表转换为 Map 再返回
static Map<String, ZoneSnapshot> createSnapshot(LoadBalancerStats lbStats) {
    Map<String, ZoneSnapshot> map = new HashMap();
    Iterator var2 = lbStats.getAvailableZones().iterator();

    while(var2.hasNext()) {
        String zone = (String)var2.next();
        ZoneSnapshot snapshot = lbStats.getZoneSnapshot(zone);
        map.put(zone, snapshot);
    }

    return map;
}
```

getAvailableZones 方法是用来筛选服务区列表的。首先，它会遍历一遍 ZoneSnapshot 哈希表。在遍历的过程中，它会做两件事情：依据 ZoneSnapshot 的实例数、实例的平均负载和实例故障率等指标将不符合标准的 ZoneSnapshot 从列表中删除，同时维护一个最坏 ZoneSnapshot 列表，当某个 ZoneSnapshot 的平均负载小于但接近全局最坏负载时，就会将该 ZoneSnapshot 加入最坏 ZoneSnapshot 列表，如果某个 ZoneSnapshot 的平均负载大于最坏负载，则清空最坏 ZoneSnapshot 列表，然后以该 ZoneSnapshot 的平均负载作为全局最坏负载，继续进行最坏 ZoneSnapshot 列表的构建。在该方法最后，如果全局最坏负载大于或等于系统设定的负载阈值，则在最坏 ZoneSnapshot 列表中随机选择一个 ZoneSnapshot，将其从列表中删除。

图 5-13 显示了 getAvailableZones 方法的筛选流程。

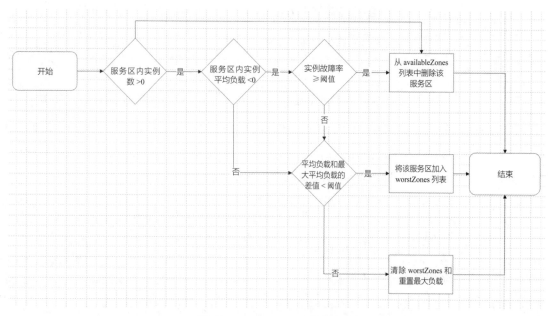

图5-13 getAvailableZones方法的筛选流程

```
//ZoneAvoidanceRule.java
public static Set<String> getAvailableZones(Map<String, ZoneSnapshot> snapshot, double
triggeringLoad, double triggeringBlackoutPercentage) {
   if (snapshot.isEmpty()) {
      return null;
   } else {
      Set<String> availableZones = new HashSet(snapshot.keySet());
      if (availableZones.size() == 1) {
         return availableZones;
      } else {
         Set<String> worstZones = new HashSet();
         double maxLoadPerServer = 0.0D;
         boolean limitedZoneAvailability = false;
         Iterator var10 = snapshot.entrySet().iterator();

// 遍历所有的服务区来判定

         while(true) {
            while(var10.hasNext()) {
               Map.Entry<String, ZoneSnapshot> zoneEntry = (Map.Entry)var10.next();
               String zone = (String)zoneEntry.getKey();
               ZoneSnapshot zoneSnapshot = (ZoneSnapshot)zoneEntry.getValue();
// 获取该服务区中的服务实例数
               int instanceCount = zoneSnapshot.getInstanceCount();
               if (instanceCount == 0) {

// 如果服务区中没有服务实例，那么去除该服务区

                  availableZones.remove(zone);
                  limitedZoneAvailability = true;
               } else {
```

```
// 如果服务区内实例平均负载小于 0 或者实例故障率（断路器断开数 / 实例数）大于等于阈值（默
认值为 0.99999），则去除该服务区

            double loadPerServer = zoneSnapshot.getLoadPerServer();
            if (!((double)zoneSnapshot.getCircuitTrippedCount() / (double)instanceCount >=
triggeringBlackoutPercentage) && !(loadPerServer < 0.0D)) {

// 如果该服务区的平均负载和最大平均负载的差值小于阈值，则将该服务区加入 worstZones 列表

                if (Math.abs(loadPerServer - maxLoadPerServer) < 1.0E-6D) {
                    worstZones.add(zone);
                } else if (loadPerServer > maxLoadPerServer) {

// 否则，如果该服务区平均负载还大于最大平均负载

                    maxLoadPerServer = loadPerServer;

// 清除 worstZones 列表，将该服务区的平均负载作为全局最坏负载

                    worstZones.clear();
                    worstZones.add(zone);
                }
            } else {
                availableZones.remove(zone);
                limitedZoneAvailability = true;
            }
        }
    }

// 如果最大平均负载小于设定的 triggeringLoad 阈值则直接返回

    if (maxLoadPerServer < triggeringLoad && !limitedZoneAvailability) {
        return availableZones;
    }

// 否则，从最坏服务区集合中随机剔除一个

    String zoneToAvoid = randomChooseZone(snapshot, worstZones);
    if (zoneToAvoid != null) {
        availableZones.remove(zoneToAvoid);
    }

    return availableZones;
    }
   }
  }
}
```

ZoneAvoidancePredicate 的 apply 方法调用结束之后，AvailabilityPredicate 的 apply 方法也会被调用。该预测规则依据断路器是否打开或者服务器连接数是否超出阈值等标准来进行服务过滤。AvailabilityPredicate 的 apply 方法如下所示。

```
// AvailabilityPredicate.java
public boolean apply(@Nullable PredicateKey input) {
```

```
        LoadBalancerStats stats = this.getLBStats();
        if (stats == null) {
            return true;
        } else {

// 获得关于该服务器的信息记录

            return !this.shouldSkipServer(stats.getSingleServerStat(input.getServer()));
        }
    }
    private boolean shouldSkipServer(ServerStats stats) {

// 如果该服务器的断路器已经打开或者服务器的连接数大于或等于预设的阈值,就需要将该服务器过滤掉

        return CIRCUIT_BREAKER_FILTERING.get() && stats.isCircuitBreakerTripped() || stats.
getActiveRequestsCount() >= (Integer)this.activeConnectionsLimit.get();
    }
```

本小节主要讲解了 ClientConfigEnabledRoundRobinRule 和 ZoneAvoidanceRule 这两个负载均衡策略,其他策略的具体实现,有兴趣的读者可以自行了解。

基于 Feign 服务调用

前面使用 RestTemplate 实现 RESTful API 调用，代码如下所示。
String url = String.format("http://%s/product/%d", "product-service", product_id);
Product product = restTemplate.getForEntity(url, Product.class).getBody();
　　由代码可知，是使用字符串占位符的方式来构造 URL 的。目前该 URL 只有两个参数。但是在现实使用中，URL 中往往含有多个参数。这时候如果还用这种方式构造 URL 就会比较麻烦。那应该如何解决呢？这个问题可以使用 Feign 来解决。
　　本章的主要内容如下。
1. 认识 Feign。
2. 使用 Feign 实现服务调用。
3. Feign 自定义配置和使用。
4. 源码解析。

6.1 认识 Feign

Feign 是 GitHub 上的一个开源项目，其目的是简化 Web Service 客户端的开发。

在使用 Feign 时，可以使用注解来修饰接口，被注解修饰的接口具有访问 Web Service 的能力，这些注解中既包括 Feign 自带的注解，也支持使用第三方的注解。除此之外，Feign 还支持插件式的编码器和解码器，使用者可以通过该特性，对请求和响应进行不同的封装与解析。

Spring Cloud 将 Feign 集成到 Netflix 项目中，与 Eureka、Ribbon 集成后，Feign 就具备了负载均衡的功能。Feign 本身在使用上的简便性，加上与 Spring Cloud 的高度整合，使用该框架在 Spring Cloud 中调用集群服务，将会大大减少开发的工作量。

6.1.1 Java 项目中接口的调用方式

Java 项目中接口的调用方式，常见的有如下几种。

（1）HttpClient。

HttpClient 是 Apache Jakarta Common 下的子项目，提供高效的、最新的、功能丰富的、支持 HTTP 的客户端编程工具包，并且支持 HTTP 最新的版本。HttpClient 相比传统 JDK 自带的 URLConnection，增加了易用性和灵活性，使客户端发送 HTTP 请求变得容易，提高了开发的效率。

（2）OkHttp。

OkHttp 是一个处理网络请求的开源项目，是安卓端最火的轻量级框架之一，由 Square 公司贡献，用于替代 HttpURLConnection 和 Apache HttpClient。OkHttp 有简洁的 API、高效的性能，并支持多种协议。

（3）HttpURLConnection。

HttpURLConnection 是 Java 的标准类，继承 URLConnection，可用于向指定网站发送 GET 请求、POST 请求。HttpURLConnection 使用比较复杂，不像 HttpClient 那样容易使用。

（4）RestTemplate。

RestTemplate 是 Spring 提供的用于访问 REST 服务的客户端，RestTemplate 提供了多种便捷访问远程 HTTP 服务的方法，能够大大提高客户端的编写效率。

上面介绍的是常见的几种调用接口的方式，下面要介绍的方式比上面的更简单、更方便，它就是 Feign。

Feign 是一个声明式的 REST 客户端，它能让 REST 调用更加简单。Feign 提供了 HTTP 请求的模板，通过编写简单的接口和插入注解，就可以定义好 HTTP 请求的参数、格式、地址等信息。Feign 会完全代理 HTTP 请求，我们只需要像调用方法一样调用它就可以完成服务请求及相关处理。Spring Cloud 对 Feign 进行了封装，使其支持 Spring MVC 标准注解和 HttpMessageConverters。Feign 可以与 Eureka 和 Ribbon 组合使用以支持负载均衡。

6.1.2 Feign 和 Ribbon 的关系

Ribbon 是一个基于 HTTP 和 TCP 客户端的负载均衡的工具。它可以在客户端配置 RibbonServerList（服务端列表），使用 HttpClient 或 RestTemplate 模拟 HTTP 请求，但步骤相当烦琐。

Feign 在 Ribbon 的基础上进行了一次改进，是一个使用起来更加方便的 HTTP 客户端。它采用接口的方式，只需要创建一个接口，然后在其中添加注解即可将需要调用的其他服务的方法定义成抽象方法，不需要自己构建 HTTP 请求，就像是调用自身工程的方法一样，感觉不到是调用远程方法，从而使得编写客户端变得非常容易。

6.2 使用 Feign 实现服务调用

以 Eureka 为注册中心，完成订单微服务调用商品微服务。为了模拟负载均衡，启动两个商品微服务，端口号分别设置为 8001 和 8002。

6.2.1 坐标依赖

在服务消费者 order-service 中添加 Feign 依赖，如下所示。

```
<dependency>
<groupId>org.springframework.cloud</groupId>
<artifactId>spring-cloud-starter-openfeign</artifactId>
</dependency>
```

6.2.2 工程改造

将贯穿案例的部分代码进行修改，目前的项目结构如图 6-1 所示。

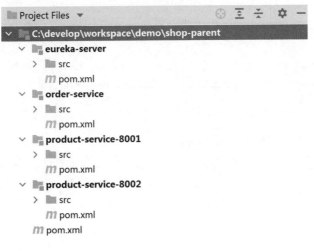

图6-1 项目结构

■ 修改启动类 order-service#OrderApplication

修改 OrderApplication 类代码如下所示。

```java
@SpringBootApplication
@EntityScan("cn.book.order.entity")
@EnableFeignClients
public class OrderApplication {

    public static void main(String[] args) {
        SpringApplication.run(OrderApplication.class, args);
    }
}
```

通过 @EnableFeignClients 注解开启 Spring Cloud Feign 的功能。

■ 激活 FeignClient

创建一个 Feign 接口，此接口是在 Feign 中调用微服务的核心接口。在服务消费者 order-service 中添加一个 ProductFeignClient 接口，代码如下所示。

```java
@FeignClient("product-service")
public interface ProductFeignClient {
    @RequestMapping(value = "/product/{id}", method = RequestMethod.GET)
    public Product findById(@PathVariable Long id);
}
```

ProductFeignClient 接口头部注解 FeignClient 中 value 的值是调用的微服务名称。order-service 要调用 product-service，接口方法 findById 头部注解 RequestMapping 中 value 的值是 product-service 中要调用的 URL 地址。简单的办法是将 product-service#ProductController#findById 方法复制过来，进行修改即可。

■ 配置请求提供者的调用接口

修改 product-service#ProductController，添加 ProductFeignClient 的自动注入，并在 order 方法中使用 ProductFeignClient 来完成微服务调用。order 方法的代码如下所示。

```java
    @RequestMapping(value = "/{product_id}/{product_num}", method = RequestMethod.GET)
    public Map<String, Object> order(@PathVariable Long product_id, @PathVariable Integer product_num) {
        // 使用 Feign 调用 HTTP 接口
        Product product = productFeignClient.findById(product_id);
        // 创建字典对象，存放信息
        Map<String, Object> map = new HashMap<String, Object>();
        // 存数据
        map.put("product_num", product_num);
        map.put("product", product);
        // 返回数据
        return map;
    }
```

6.2.3 代码测试

重启各个服务后，访问注册中心管理后台，结果如图 6-2 所示。

10.211.55.12:9000
spring Eureka
HOME
LAST 1000 SINCE STARTUP
System Status
Environment
test
Data center
default
Current time
2021-02-02T09:11:39 +0800
Uptime
00:32
Lease expiration enabled
true
Renews threshold
0
Renews (last min)
6
THE SELF PRESERVATION MODE IS TURNED OFF. THIS MAY NOT PROTECT INSTANCE EXPIRY IN CASE OF NETWORK/OTHER PROBLEMS.
DS Replicas
Instances currently registered with Eureka
Application
AMIs
Availability Zones
Status
ORDER-SERVICE
n/a (1)
(1)
UP (1) - 10.211.55.12:9003
PRODUCT-SERVICE
n/a (2)
(2)
UP (2) - 10.211.55.12:8001 , 10.211.55.12:8002

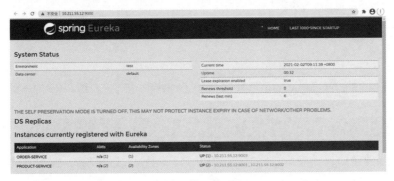

图6-2 注册中心管理后台

微服务 PRODUCT-SERVICE 对应的 Status 下已经显示了添加的两个地址。

访问订单微服务，结果如图 6-3 所示。

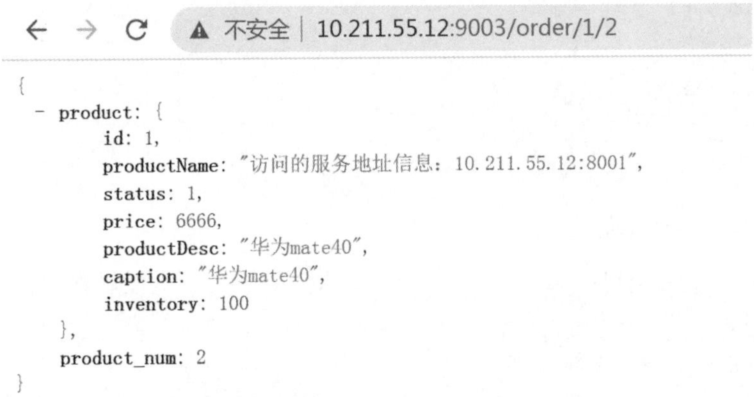

不安全 10.211.55.12:9003/order/1/2
{
- product: {
id: 1,
productName: "访问的服务地址信息：10.211.55.12:8001",
status: 1,
price: 6666,
productDesc: "华为mate40",
caption: "华为mate40",
inventory: 100
},
product_num: 2
}

图6-3 访问订单微服务

再次访问订单微服务，结果如图 6-4 所示。

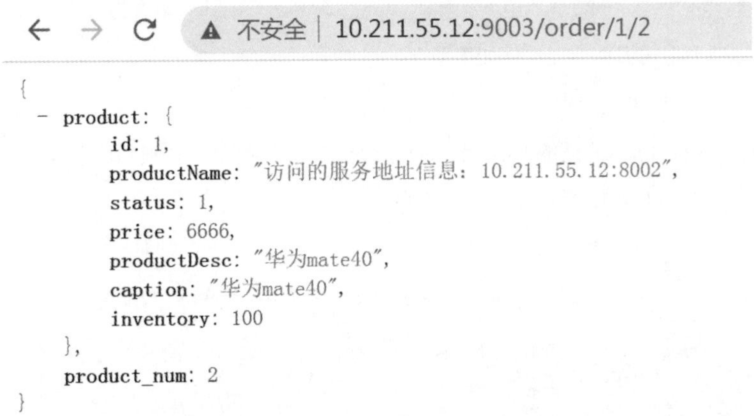

不安全 10.211.55.12:9003/order/1/2
{
- product: {
id: 1,
productName: "访问的服务地址信息：10.211.55.12:8002",
status: 1,
price: 6666,
productDesc: "华为mate40",
caption: "华为mate40",
inventory: 100
},
product_num: 2
}

图6-4 再次访问订单微服务

通过两次访问订单微服务可以发现，订单微服务已按轮询的方式调用了商品微服务。Feign 服务调用和负载均衡测试完成。

Feign 自定义配置和使用

Feign 提供了很多的扩展机制，让用户可以更加灵活地使用。接下来介绍 Feign 的一些自定义配置。

6.3.1 日志配置

有时候使用者遇到 bug（漏洞），比如接口调用失败、参数没收到等问题，或者想看看调用性能，就需要配置 Feign 的日志，从而让 Feign 把请求信息输出来。首先定义一个配置类，代码如下所示。

```
@Configuration
public class FeignConfiguration {
    /**
     * 日志级别
     */
    @Bean
    Logger.Level feignLoggerLevel() {
        return Logger.Level.FULL;
    }
}
```

Feign 日志级别源码如下所示。

```
public static enum Level {
    NONE,
    BASIC,
    HEADERS,
    FULL;

    private Level() {
    }
}
```

通过源码可以看到日志级别有 4 种，如下所示。

（1）NONE：不输出日志。

（2）BASIC：只输出请求方法的 URL 和响应的状态码以及接口执行的时间。

（3）HEADERS：将 BASIC 信息和请求头信息输出。

（4）FULL：输出完整的请求信息。

配置类建好后，需要在 FeignClient 中的 @FeignClient 注解中指定使用的配置类，代码如下所示。

```
@FeignClient(value = "product-service",configuration = FeignConfiguration.class)
public interface ProductFeignClient {
    @RequestMapping(value = "/product/{id}", method = RequestMethod.GET)
    public Product findById(@PathVariable Long id);
}
```

还需要修改 application.yml 文件，设置日志输出级别，如下所示。

```yaml
server:
  port: 9003 # 端口
spring:
  application:
    name: order-service # 服务名称
  datasource: # 数据源
    driver-class-name: com.mysql.jdbc.Driver
    url: jdbc:mysql://192.168.10.167:3306/shop?characterEncoding=utf8
    username: root
    password: root
  jpa:
    database: MySQL
    show-sql: true
    open-in-view: true
eureka: # 配置 Eureka
  client:
    service-url:
      defaultZone: http://10.211.55.12:9000/eureka/
  instance:
    prefer-ip-address: true # 使用 IP 地址注册
    instance-id: ${spring.cloud.client.ip-address}:${server.port}
logging: # 设置日志级别
  level:
    cn.book:
      DEBUG
```

重启订单微服务，访问商品微服务，查看订单微服务的控制台，日志信息如下所示。

```
    2021-02-03 18:39:44.320 DEBUG 3220 --- [nio-9003-exec-8] cn.book.order.feign.ProductFeignClient    : [ProductFeignClient#findById] ---> GET http://product-service/product/1 HTTP/1.1
    2021-02-03 18:39:44.322 DEBUG 3220 --- [nio-9003-exec-8] cn.book.order.feign.ProductFeignClient    : [ProductFeignClient#findById] ---> END HTTP (0-byte body)
    2021-02-03 18:39:44.644 DEBUG 3220 --- [nio-9003-exec-8] cn.book.order.feign.ProductFeignClient    : [ProductFeignClient#findById] <--- HTTP/1.1 200 (321ms)
    2021-02-03 18:39:44.644 DEBUG 3220 --- [nio-9003-exec-8] cn.book.order.feign.ProductFeignClient    : [ProductFeignClient#findById] content-type: application/json;charset=UTF-8
    2021-02-03 18:39:44.644 DEBUG 3220 --- [nio-9003-exec-8] cn.book.order.feign.ProductFeignClient    : [ProductFeignClient#findById] date: Wed, 03 Feb 2021 10:39:44 GMT
    2021-02-03 18:39:44.644 DEBUG 3220 --- [nio-9003-exec-8] cn.book.order.feign.ProductFeignClient    : [ProductFeignClient#findById] transfer-encoding: chunked
    2021-02-03 18:39:44.644 DEBUG 3220 --- [nio-9003-exec-8] cn.book.order.feign.ProductFeignClient    : [ProductFeignClient#findById]
    2021-02-03 18:39:44.647 DEBUG 3220 --- [nio-9003-exec-8] cn.book.order.feign.ProductFeignClient    : [ProductFeignClient#findById] {"id":1,"productName":"访问的服务地址信息：10.211.55.12:8002","status":1,"price":6666.00,"productDesc":"华为mate40","caption":"华为mate40","inventory":100}
    2021-02-03 18:39:44.647 DEBUG 3220 --- [nio-9003-exec-8] cn.book.order.feign.ProductFeignClient    : [ProductFeignClient#findById] <--- END HTTP (169-byte body)
    2021-02-03 18:43:49.808  INFO 3220 --- [trap-executor-0] c.n.d.s.r.aws.ConfigClusterResolver      : Resolving eureka endpoints via configuration
```

6.3.2 超时时间配置

通过 Options 可以配置连接超时时间和读取超时时间，Options 的第一个参数是连接超时时间（单位为 ms），默认值是 10×1000；第二个参数是读取超时时间（单位为 ms），默认值是 60×1000，如下所示。

```
@Configuration
public class FeignConfiguration {
  /**
   * 日志级别
   */
  @Bean
  Logger.Level feignLoggerLevel() {
    return Logger.Level.FULL;
  }

  /**
   * 超时时间
   */
  @Bean
  public Request.Options options() {
    return new Request.Options(5000, 10000);
  }
}
```

6.3.3 客户端组件配置

Feign 默认使用 JDK 原生的 URLConnection 发送 HTTP 请求，使用者可以集成别的组件来替换 URLConnection，比如 Apache HttpClient 或 OkHttp。

配置 OkHttp 只需要加入 OkHttp 的依赖，代码如下所示。

```xml
<dependency>
    <groupId>io.github.openfeign</groupId>
    <artifactId>feign-okhttp</artifactId>
</dependency>
```

然后修改配置，将 Feign 的 HttpClient 禁用，启用 OkHttp。相关配置如下：

```
feign: #Feign 配置
  httpclient:
    enabled: false
  okhttp:
    enabled: true
```

关于配置可参考源码 org.springframework.cloud.openfeign.FeignAutoConfiguration。

HttpClient 自动配置源码如下所示。

```
@Configuration
@ConditionalOnClass(ApacheHttpClient.class)
@ConditionalOnMissingClass("com.netflix.loadbalancer.ILoadBalancer")
@ConditionalOnProperty(value = "feign.httpclient.enabled", matchIfMissing = true)
protected static class HttpClientFeignConfiguration {
```

```
    @Autowired(required = false)
    private HttpClient httpClient;
    @Bean
    @ConditionalOnMissingBean(Client.class)
    public Client feignClient() {
        if (this.httpClient != null) {
            return new ApacheHttpClient(this.httpClient);
        }
        return new ApacheHttpClient();
    }
}
```

OkHttp 自动配置源码如下所示。

```
@Configuration
@ConditionalOnClass(OkHttpClient.class)
@ConditionalOnMissingClass("com.netflix.loadbalancer.ILoadBalancer")
@ConditionalOnProperty(value = "feign.okhttp.enabled", matchIfMissing = true)
protected static class OkHttpFeignConfiguration {
    @Autowired(required = false)
    private okhttp3.OkHttpClient okHttpClient;
    @Bean
    @ConditionalOnMissingBean(Client.class)
    public Client feignClient() {
        if (this.okHttpClient != null) {
            return new OkHttpClient(this.okHttpClient);
        }
        return new OkHttpClient();
    }
}
```

上面所示两段代码分别是配置 HttpClient 和 OkHttp 的方法，通过 @ConditionalOnProperty 中的值来决定启用哪种客户端（HttpClient 还是 OkHttp），@ConditionalOnClass 表示只有对应的类在 classpath 目录下存在时，才会去解析对应的配置文件。

6.3.4 压缩配置

开启压缩可以有效节约网络资源，提升接口性能，可以通过配置 GZIP 来压缩数据，如下所示。

```
feign:
  httpclient:
    enabled: false
  okhttp:
    enabled: true
  compression:
    request:
      enabled: true
    response:
      enabled: true
```

还可以配置压缩的类型、最小压缩值的标准，如下所示。

```
feign:
```

```yaml
httpclient:
  enabled: false
okhttp:
  enabled: true
compression:
  request:
    enabled: true
    mime-types: text/xml,application/xml,application/json
    min-request-size: 2048
  response:
    enabled: true
```

只有当启用的客户端不是 OkHttp 的时候，压缩才会生效。配置源码在 org.springframework.cloud.openfeign.encoding.FeignAcceptGzipEncodingAutoConfiguration 中，代码如下所示。

```java
@Configuration
@EnableConfigurationProperties(FeignClientEncodingProperties.class)
@ConditionalOnClass(Feign.class)
@ConditionalOnBean(Client.class)
@ConditionalOnProperty(value = "feign.compression.response.enabled", matchIfMissing = false)
@ConditionalOnMissingBean(type = "okhttp3.OkHttpClient")
@AutoConfigureAfter(FeignAutoConfiguration.class)
public class FeignAcceptGzipEncodingAutoConfiguration {
    @Bean
    public FeignAcceptGzipEncodingInterceptor feignAcceptGzipEncodingInterceptor(
        FeignClientEncodingProperties properties) {
        return new FeignAcceptGzipEncodingInterceptor(properties);
    }
}
```

核心代码就是 @ConditionalOnMissingBean(type="okhttp3.OkHttpClient")，表示 Spring BeanFactory 中不包含指定的 Bean 条件时匹配，也就是没有启用 OkHttp 时才会进行压缩配置。

6.3.5 使用配置文件自定义 Feign 的配置

除了使用代码的方式对 Feign 进行配置，还可以通过配置文件来自定义 Feign 的配置。

```
# 连接超时时间
feign.client.config.feignName.connectTimeout=5000
# 读取超时时间
feign.client.config.feignName.readTimeout=5000
# 日志级别
feign.client.config.feignName.loggerLevel=full
# 重试
feign.client.config.feignName.retryer=com.example.SimpleRetryer
# 拦截器
feign.client.config.feignName.requestInterceptors[0]=com.example.FooRequestInterceptor
feign.client.config.feignName.requestInterceptors[1]=com.example.BarRequestInterceptor
# 编码器
feign.client.config.feignName.encoder=com.example.SimpleEncoder
# 解码器
feign.client.config.feignName.decoder=com.example.SimpleDecoder
# 契约
```

feign.client.config.feignName.contract=com.example.SimpleContract

 ## 6.4 源码分析

读 OpenFeign 源码时，可以沿着两条路线进行，一是 FeignServiceClient 等被 @FeignClient 注解修饰的接口类（后续简称为 FeignClient 接口类）是如何创建的，也就是 Bean 实例是如何被创建的；二是调用 FeignServiceClient 对象的网络请求相关的函数时，OpenFeign 是如何发送网络请求的。OpenFeign 相关的类也可以以此来进行分类，一部分是用来初始化相应的 Bean 实例的，另一部分是用来在调用方法时发送网络请求的。

6.4.1 核心组件与概念

图 6-5 所示是 OpenFeign 相关的关键类图，其中比较重要的类是 FeignClientFactoryBean、FeignContext 和 SynchronousMethodHandler。FeignClientFactoryBean 用于创建 FeignClient 接口类 Bean 实例的工厂类；FeignContext 用于配置组件的上下文环境，保存着相关组件的不同实例，这些实例是由不同的 FeignConfiguration 类构造出来的；SynchronousMethodHandler 是 MethodHandler 的子类，可以在 FeignClient 相应方法被调用时发送网络请求，然后将请求响应转化为函数返回值进行输出。

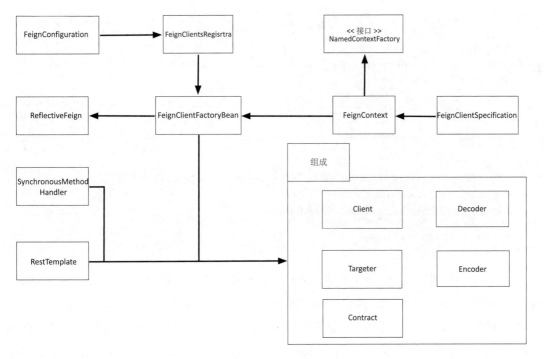

图6-5　OpenFeign相关的关键类图

6.4.2 动态注册 BeanDefinition

如图 6-6 所示，OpenFeign 首先进行相关 BeanDefinition 的动态注册，然后当 Spring 容器注入相关实例时进行实例的初始化，最后当 FeignClient 接口类实例函数被调用时发送网络请求。

图6-6　源码流程

OpenFeign 可以通过多种方式进行自定义配置，配置的变化会导致接口类初始化时使用不同的 Bean 实例，从而控制 OpenFeign 的相关行为，比如网络请求的编解码、压缩和日志处理。可以说，了解 OpenFeign 配置和实例初始化的流程与原理对于使用者学习和使用 OpenFeign 有着至关重要的作用，而且 Spring Cloud 所有项目的配置和实例初始化过程的原理基本相同，了解了 OpenFeign 的原理，就可以触类旁通。

■ FeignClientsRegistrar

@EnableFeignClients 就像 OpenFeign 的开关一样，一切 OpenFeign 的相关操作都是从它开始的。@EnableFeignClients 有 3 个功能，一是引入 FeignClientsRegistrar；二是指定扫描 FeignClient 的包信息，就是指定 FeignClient 接口类所在的包名；三是指定 FeignClient 接口类的自定义配置类。@EnableFeignClients 注解的定义如下所示。

```
@Retention(RetentionPolicy.RUNTIME)
@Target({ElementType.TYPE})
@Documented
@Import({FeignClientsRegistrar.class})
public @interface EnableFeignClients {
// 下面 3 个函数都是为了指定需要扫描的包
    String[] value() default {};

    String[] basePackages() default {};

    Class<?>[] basePackageClasses() default {};
// 自定义 FeignClient 的配置，可以配置 Decoder、Encoder 和 Contract 等组件
// FeignClientsConfiguration 是默认的配置类
    Class<?>[] defaultConfiguration() default {};
// 指定被 @FeignClient 修饰的类，如果不为空，那么路径自动检测机制会被关闭
    Class<?>[] clients() default {};
}
```

上面的代码中，FeignClientsRegistrar 是 ImportBeanDefinitionRegistrar 的子类，Spring 用 ImportBeanDefinitionRegistrar 来动态注册 BeanDefinition。OpenFeign 通过 FeignClientsRegistrar 来处理 @FeignClient 修饰的 FeignClient 接口类，将这些接口类的 BeanDefinition 注册到 Spring 容器中，这样就可以使用 @Autowired 等方式来自动装载这些 FeignClient 接口类的 Bean 实例。FeignClientsRegistrar 的部分代码如下所示。

```java
class FeignClientsRegistrar implements ImportBeanDefinitionRegistrar, ResourceLoaderAware,
EnvironmentAware {
   ......
   public void registerBeanDefinitions(AnnotationMetadata metadata, BeanDefinitionRegistry registry) {
   // 根据 EnableFeignClients 的属性值来构建 Feign 的自定义配置类进行注册
   this.registerDefaultConfiguration(metadata, registry);
   // 扫描包，注册被 @FeignClient 修饰的接口类的 Bean 实例
   this.registerFeignClients(metadata, registry);
   }
   ......
```

如上述代码所示，FeignClientsRegistrar 的 registerBeanDefinitions 方法主要做了两件事情，一是注册 @EnableFeignClients 提供的自定义配置类中的相关 Bean 实例，二是根据 @EnableFeignClients 提供的包信息扫描 FeignClient 接口类，然后进行 Bean 实例注册。

@EnableFeignClients 的自定义配置类是被 @Configuration 注解修饰的配置类，它会提供组装 FeignClient 的各类组件实例。这些组件包括 Client、Targeter、Decoder、Encoder 和 Contract 等。接下来看看 registerDefaultConfiguration 方法的代码实现，如下所示。

```java
   private void registerDefaultConfiguration(AnnotationMetadata metadata, BeanDefinitionRegistry
registry) {
// 获取 metadata 中关于 EnableFeignClients 的属性键值对
      Map<String, Object> defaultAttrs = metadata.getAnnotationAttributes(EnableFeignClients.
class.getName(), true);
// 如果 EnableFeignClients 配置了 defaultConfiguration 类，那么进行下一步操作
// 否则，使用默认的 FeignConfiguration
   if (defaultAttrs != null && defaultAttrs.containsKey("defaultConfiguration")) {
         String name;
         if (metadata.hasEnclosingClass()) {
            name = "default." + metadata.getEnclosingClassName();
         } else {
            name = "default." + metadata.getClassName();
         }

         this.registerClientConfiguration(registry, name, defaultAttrs.get("defaultConfiguration"));
      }

   }
```

如上述代码所示，registerDefaultConfiguration 方法会判断 @EnableFeignClients 注解是否设置了 defaultConfiguration 属性。如果有设置，则调用 registerClientConfiguration 方法进行 BeanDefinitionRegistry 的注册。registerClientConfiguration 方法的代码如下所示。

```java
   private void registerClientConfiguration(BeanDefinitionRegistry registry, Object name, Object
configuration) {
// 使用 BeanDefinitionBuilder 生成 BeanDefinition，并注册到 registry 上
      BeanDefinitionBuilder builder = BeanDefinitionBuilder.genericBeanDefinition(FeignClientSpecifi
cation.class);
      builder.addConstructorArgValue(name);
      builder.addConstructorArgValue(configuration);
      registry.registerBeanDefinition(name + "." + FeignClientSpecification.class.getSimpleName(),
```

```
        builder.getBeanDefinition());
    }
```

BeanDefinitionRegistry 是 Spring 框架中用于动态注册 BeanDefinition 的接口，调用其自身的 registerBeanDefinition 方法可以将 BeanDefinition 注册到 Spring 容器中，其中 name 属性就是注册的 BeanDefinition 的名称。

FeignClientSpecification 类实现了 NamedContextFactory.Specification 接口，它是 OpenFeign 组件实例化的重要一环，持有自定义配置类提供的组件实例，供 OpenFeign 使用。Spring Cloud 使用 NamedContextFactory 创建一系列的运行上下文 ApplicationContext 来让对应的 Specification 在这些上下文中创建实例对象。这样使得各个子上下文中的实例对象相互独立、互不影响，可以方便地通过子上下文管理一系列不同的实例对象。

NamedContextFactory 有 3 个功能，一是创建子上下文 AnnotationConfigApplicationContext；二是在子上下文中创建并获取 Bean 实例；三是当子上下文消亡时清除其中的 Bean 实例。在 OpenFeign 中，FeignContext 继承了 NamedContextFactory，用于存储 OpenFeign 组件的实例。图 6-7 所示就是 FeignContext 的相关类图。

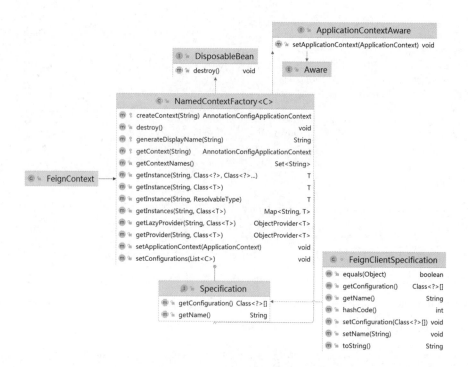

图6-7　FeignContext的相关类图

FeignAutoConfiguration 是 OpenFeign 的自动配置类。它会提供 FeignContext 实例，并且将之前注册的 FeignClientSpecification 通过 setConfigurations 方法设置给 FeignContext 实例。同时处理了默认配置类 FeignClientsConfiguration 和自定义配置类的替换问题。如果 FeignClientsRegistrar 没有注册自定义配置类，

那么 configurations 将不包含 FeignClientSpecification 对象，否则会在 setConfigurations 方法中进行默认配置类的替换。FeignAutoConfiguration 的相关代码如下所示。

```java
//FeignAutoConfiguration.java
@Autowired(
    required = false
)
private List<FeignClientSpecification> configurations = new ArrayList();

@Bean
public FeignContext feignContext() {
    FeignContext context = new FeignContext();
    context.setConfigurations(this.configurations);
    return context;
}

// FeignContext.java
public class FeignContext extends NamedContextFactory<FeignClientSpecification> {
    public FeignContext() {
// 将默认的 FeignClientsConfiguration 作为参数传递给构造方法
        super(FeignClientsConfiguration.class, "feign", "feign.client.name");
    }
}
```

NamedContextFactory 是 FeignContext 的父类。Named ContextFactory 的 createContext 方法会创建具有名称的 Spring 的 AnnotationConfigApplicationContext 实例作为当前上下文的子上下文。这些 AnnotationConfigApplicationContext 实例可以管理 OpenFeign 组件的不同实例。NamedContextFactory 的实现代码如下所示。

```java
//NamedContextFactory.java
protected AnnotationConfigApplicationContext createContext(String name) {
    AnnotationConfigApplicationContext context = new AnnotationConfigApplicationContext();
// 获取该 name 所对应的 configuration, 如果有，就注册到子上下文中
    if (this.configurations.containsKey(name)) {
        Class[] var3 = ((NamedContextFactory.Specification)this.configurations.get(name)).getConfiguration();
        int var4 = var3.length;

        for(int var5 = 0; var5 < var4; ++var5) {
            Class<?> configuration = var3[var5];
            context.register(new Class[]{configuration});
        }
    }

    Iterator var9 = this.configurations.entrySet().iterator();

    while(true) {
        Entry entry;
        do {
            if (!var9.hasNext()) {
                    context.register(new Class[]{PropertyPlaceholderAutoConfiguration.class, this.defaultConfigType});
                    context.getEnvironment().getPropertySources().addFirst(new MapPropertySource(this.
```

```
            propertySourceName, Collections.singletonMap(this.propertyName, name)));
              if (this.parent != null) {
                 context.setParent(this.parent);
              }

              context.setDisplayName(this.generateDisplayName(name));
              context.refresh();
              return context;
           }

           entry = (Entry)var9.next();
       } while(!((String)entry.getKey()).startsWith("default."));

       Class[] var11 = ((NamedContextFactory.Specification)entry.getValue()).getConfiguration();
       int var12 = var11.length;

       for(int var7 = 0; var7 < var12; ++var7) {
          Class<?> configuration = var11[var7];
          context.register(new Class[]{configuration});
       }
    }
}
```

而由于 NamedContextFactory 实现了 DisposableBean 接口,当 NamedContextFactory 实例消亡时,Spring 框架会调用 NamedContextFactory 的 destroy 的方法,清除 NamedContextFactory 创建的所有子上下文和自身包含的所有组件实例。NamedContextFactory 的 destroy 方法如下所示。

```
//NamedContextFactory.java
public void destroy() {
    Collection<AnnotationConfigApplicationContext> values = this.contexts.values();
    Iterator var2 = values.iterator();

    while(var2.hasNext()) {
        AnnotationConfigApplicationContext context = (AnnotationConfigApplicationContext)var2.next();
        context.close();
    }

    this.contexts.clear();
}
```

NamedContextFactory 会创建出 AnnotationConfigApplicationContext 实例,并以 name 作为唯一标识,然后每个 AnnotationConfigApplicationContext 实例都会注册部分配置类,从而可以给出一系列的基于配置类生成的组件实例,这样就可以基于 name 来管理一系列的组件实例,为不同的 FeignClient 准备不同配置组件,比如 Decoder、Encoder 等。

■ 扫描类信息

FeignClientsRegistrar 还会扫描指定包下的类文件,注册 FeignClient 接口类,如下所示。

```
// FeignClientsRegistrar.java
public void registerFeignClients(AnnotationMetadata metadata, BeanDefinitionRegistry registry) {
```

```java
// 生成自定义的 ClassPathScanning CandidateComponent Provider
    ClassPathScanningCandidateComponentProvider scanner = this.getScanner();
    scanner.setResourceLoader(this.resourceLoader);

// 获取 EnableFeignClients 所有属性的键值对
        Map<String, Object> attrs = metadata.getAnnotationAttributes(EnableFeignClients.class.getName());

// 根据 Annotation 来进行 TypeFilter，只会扫描出 FeignClient 接口类
    AnnotationTypeFilter annotationTypeFilter = new AnnotationTypeFilter(FeignClient.class);
    Class<?>[] clients = attrs == null ? null : (Class[])((Class[])attrs.get("clients"));
    Object basePackages;

// 如果没有设置 clients 属性，那么需要扫描 basePackages
// 如果设置了 annotationTypeFilter 就获取 basePackages
if (clients != null && clients.length != 0) {
        final Set<String> clientClasses = new HashSet();
        basePackages = new HashSet();
        Class[] var9 = clients;
        int var10 = clients.length;

        for(int var11 = 0; var11 < var10; ++var11) {
            Class<?> clazz = var9[var11];
            ((Set)basePackages).add(ClassUtils.getPackageName(clazz));
            clientClasses.add(clazz.getCanonicalName());
        }

        AbstractClassTestingTypeFilter filter = new AbstractClassTestingTypeFilter() {
            protected boolean match(ClassMetadata metadata) {
                String cleaned = metadata.getClassName().replaceAll("\\$", ".");
                return clientClasses.contains(cleaned);
            }
        };
        scanner.addIncludeFilter(new FeignClientsRegistrar.AllTypeFilter(Arrays.asList(filter, annotationTypeFilter)));
    } else {
        scanner.addIncludeFilter(annotationTypeFilter);
        basePackages = this.getBasePackages(metadata);
    }

    Iterator var17 = ((Set)basePackages).iterator();

    while(var17.hasNext()) {
        String basePackage = (String)var17.next();
        Set<BeanDefinition> candidateComponents = scanner.findCandidateComponents(basePackage);
        Iterator var21 = candidateComponents.iterator();

        while(var21.hasNext()) {
            BeanDefinition candidateComponent = (BeanDefinition)var21.next();
            if (candidateComponent instanceof AnnotatedBeanDefinition) {
                AnnotatedBeanDefinition beanDefinition = (AnnotatedBeanDefinition)candidateComponent;
                AnnotationMetadata annotationMetadata = beanDefinition.getMetadata();
```

```java
            Assert.isTrue(annotationMetadata.isInterface(), "@FeignClient can only be specified on an interface");
            Map<String, Object> attributes = annotationMetadata.getAnnotationAttributes(FeignClient.class.getCanonicalName());

            String name = this.getClientName(attributes);
            this.registerClientConfiguration(registry, name, attributes.get("configuration"));

// 注册 FeignClient 的 BeanDefinition
            this.registerFeignClient(registry, annotationMetadata, attributes);
        }
    }
}
```

如上述代码所示,FeignClientsRegistrar 的 registerFeignClients 方法根据 @EnableFeignClients 的属性获取要扫描的包信息,然后获取这些包下所有 FeignClient 接口类的 BeanDefinition,最后调用 registerFeignClient 动态注册 BeanDefinition。

registerFeignClients 方法中有一些细节值得认真学习,有利于加深对 Spring 框架的理解。首先是如何自定义 Spring 类扫描器,即如何使用 ClassPathScanningCandidateComponentProvider 和各种 TypeFilter。

OpenFeign 使用了 AnnotationTypeFilter 来过滤 FeignClient 接口类。

getScanner 方法的具体实现如下所示。

```java
// FeignClientsRegistrar.java
protected ClassPathScanningCandidateComponentProvider getScanner() {
    return new ClassPathScanningCandidateComponentProvider(false, this.environment) {
        protected boolean isCandidateComponent(AnnotatedBeanDefinition beanDefinition) {
            boolean isCandidate = false;

// 判断 beanDefinition 是否为内部类
// 判断是否为接口类,并且该接口是否是 Annotation
            if (beanDefinition.getMetadata().isIndependent() && !beanDefinition.getMetadata().isAnnotation()) {
                isCandidate = true;
            }

            return isCandidate;
        }
    };
}
```

ClassPathScanningCandidateComponentProvider 的作用是遍历指定路径的包下的所有类。比如指定包路径为 com/test/openfeign,它会找出 com.test.openfeign 包下所有的类,将所有的类封装成 Resource 接口集合。Resource 接口是 Spring 对资源的封装,有 FileSystemResource、ClassPathResource、UrlResource 等多种实现。接着 ClassPathScanningCandidateComponentProvider 类会遍历 Resource 接口集合,通过 includeFilters 和 excludeFilters 两种过滤器进行过滤操作。includeFilters 和 excludeFilters 是 TypeFilter 接口类型实例的集合,TypeFilter 接口是用于判断类型是否满足要求的类型过滤器。excludeFilters 中只要有一个 TypeFilter 满足条

件，这个 Resource 接口就会被过滤；而 includeFilters 中只要有一个 TypeFilter 满足条件，这个 Resource 接口就不会被过滤。如果一个 Resource 接口没有被过滤，就会被转换成 ScannedGenericBeanDefinition 添加到 BeanDefinition 集合中。

6.4.3 实例初始化

FeignClientFactoryBean 是工厂类，Spring 容器通过调用它的 getObject 方法来获取对应的 Bean 实例。FeignClient 接口类都是通过 FeignClientFactoryBean 的 getObject 方法来进行实例化的，具体实现代码如下所示。

```java
// FeignClientFactoryBean.java
public Object getObject() throws Exception {
    return this.getTarget();
}

<T> T getTarget() {
    FeignContext context = (FeignContext)this.applicationContext.getBean(FeignContext.class);
    Builder builder = this.feign(context);
    if (!StringUtils.hasText(this.url)) {
        if (!this.name.startsWith("http")) {
            this.url = "http://" + this.name;
        } else {
            this.url = this.name;
        }

        this.url = this.url + this.cleanPath();
        return this.loadBalance(builder, context, new HardCodedTarget(this.type, this.name, this.url));
    } else {
        if (StringUtils.hasText(this.url) && !this.url.startsWith("http")) {
            this.url = "http://" + this.url;
        }

        String url = this.url + this.cleanPath();

// 调用 FeignContext 的 getInstance 方法获取 Client 对象
        Client client = (Client)this.getOptional(context, Client.class);

// 因为有具体的 URL，不需要负载均衡，所以去除 LoadBalancerFeignClient 实例
        if (client != null) {
            if (client instanceof LoadBalancerFeignClient) {
                client = ((LoadBalancerFeignClient)client).getDelegate();
            }

            builder.client(client);
        }

        Targeter targeter = (Targeter)this.get(context, Targeter.class);
        return targeter.target(this, builder, context, new HardCodedTarget(this.type, this.name, url));
    }
}
```

上述代码就用到了 FeignContext 的 getInstance 方法，getOptional 方法调用了 FeignContext 的 getInstance 方法，从 FeignContext 的对应名称的子上下文中获取 Client 类型的 Bean 实例，其具体实现如下所示。

```java
// NamedContextFactory.java
public <T> T getInstance(String name, Class<T> type) {
    AnnotationConfigApplicationContext context = this.getContext(name);

// 从对应的 context 中获取 Bean 实例，如果对应的子上下文中没有则直接从父上下文中获取
    return BeanFactoryUtils.beanNamesForTypeIncludingAncestors(context, type).length > 0 ?
    context.getBean(type) : null;
}
```

默认情况下，子上下文中并没有这些类型的 BeanDefinition，只能从父上下文中获取，而父上下文中 Client 类型的 BeanDefinition 是在 FeignAutoConfiguration 中注册的。但是如果子上下文注册的配置类提供了 Client 类型的 Bean 实例，子上下文会直接将自己配置类的 Client 类型的 Bean 实例返回，否则由父上下文返回默认 Client 类型的 Bean 实例。Client 在 FeignAutoConiguration 中的配置如下所示。

```java
// FeignAutoConfiguration.java
@ConditionalOnMissingBean({Client.class})
public Client feignClient(okhttp3.OkHttpClient client) {
return new OkHttpClient(client);
}
```

Targeter 是一个接口，它的 target 方法会生成对应的实例对象。Targeter 有两个实现类，分别为 DefaultTargeter 和 HystrixTargeter。OpenFeign 使用 HystrixTargeter 来封装关于 Hystrix 的实现。DefaultTargeter 的实现只是调用了 Feign.Builder 的 target 方法，如下所示。

```java
// DefaultTargeter.java
class DefaultTargeter implements Targeter {
    DefaultTargeter() {
    }

    public <T> T target(FeignClientFactoryBean factory, Builder feign, FeignContext context, HardCodedTarget<T> target) {
        return feign.target(target);
    }
}
```

Feign.Builder 是由 FeignClientFactoryBean 对象的 Feign 方法创建的。Feign.Builder 会设置 FeignLoggerFactory、Encoder、Decoder 和 Contract 等组件，这些组件的 Bean 实例都是通过 FeignContext 获取的，也就是说这些实例都是可配置的，可以通过 OpenFeign 的配置机制为不同的 FeignClient 配置不同的组件实例。feign 方法的实现如下所示。

```java
// FeignClientFactoryBean.java
protected Builder feign(FeignContext context) {
        FeignLoggerFactory loggerFactory = (FeignLoggerFactory)this.get(context, FeignLoggerFactory.class);
    Logger logger = loggerFactory.create(this.type);
        Builder builder = ((Builder)this.get(context, Builder.class)).logger(logger).encoder((Encoder)this.get(context, Encoder.class)).decoder((Decoder)this.get(context, Decoder.class)).
```

```
        contract((Contract)this.get(context, Contract.class));
    this.configureFeign(context, builder);
    return builder;
}
```

Feign.Builder 负责生成 FeignClient 接口类实例。它通过 Java 反射机制构造 InvocationHandler 实例并将其注册到 FeignClient 上，当 FeignClient 的方法被调用时，InvocationHandler 的回调函数会被调用，OpenFeign 会在其回调函数中发送网络请求。实现方法如下所示。

```
// Feign.Builder
public Feign build() {
    Factory synchronousMethodHandlerFactory = new Factory(this.client, this.retryer, this.requestInterceptors, this.logger, this.logLevel, this.decode404, this.closeAfterDecode, this.propagationPolicy);
    ParseHandlersByName handlersByName = new ParseHandlersByName(this.contract, this.options, this.encoder, this.decoder, this.queryMapEncoder, this.errorDecoder, synchronousMethodHandlerFactory);
    return new ReflectiveFeign(handlersByName, this.invocationHandlerFactory, this.queryMapEncoder);
}
```

ReflectiveFeign 的 newInstance 方法是生成 FeignClient 实例的关键。它主要做两件事情，一是扫描 FeignClient 接口类的所有函数，生成对应的 Handler；二是使用 Proxy 生成 FeignClient 的实例对象。相关代码如下所示。

```
// ReflectiveFeign.java
  public <T> T newInstance(Target<T> target) {
      Map<String, MethodHandler> nameToHandler = this.targetToHandlersByName.apply(target);
      Map<Method, MethodHandler> methodToHandler = new LinkedHashMap();
      List<DefaultMethodHandler> defaultMethodHandlers = new LinkedList();
      Method[] var5 = target.type().getMethods();
      int var6 = var5.length;

      for(int var7 = 0; var7 < var6; ++var7) {
          Method method = var5[var7];
          if (method.getDeclaringClass() != Object.class) {
              if (Util.isDefault(method)) {

// 为每个默认方法生成一个 DefaultMethodHandler
                  DefaultMethodHandler handler = new DefaultMethodHandler(method);
                  defaultMethodHandlers.add(handler);
                  methodToHandler.put(method, handler);
              } else {
                  methodToHandler.put(method, (MethodHandler)nameToHandler.get(Feign.configKey(target.type(), method)));
              }
          }
      }

// 生成 Java Reflective 的 InvocationHandler
      InvocationHandler handler = this.factory.create(target, methodToHandler);
      T proxy = Proxy.newProxyInstance(target.type().getClassLoader(), new Class[]{target.type()}, handler);
```

```
        Iterator var12 = defaultMethodHandlers.iterator();

        while(var12.hasNext()) {
            DefaultMethodHandler defaultMethodHandler = (DefaultMethodHandler)var12.next();

// 将 defaultMethodHandler 绑定到 proxy 中
            defaultMethodHandler.bindTo(proxy);
        }

        return proxy;
    }
```

6.4.4 函数调用和网络请求

在配置和实例生成结束之后,就可以直接使用 FeignClient 接口类的实例,调用它的函数来发送网络请求。在调用其函数的过程中,由于设置了 MethodHandler,因此函数调用最终会执行 SynchronousMethodHandler 的 invoke 方法。在该方法中,OpenFeign 会将函数的实际参数值与之前生成的 RequestTemplate 结合,然后发送网络请求。

图 6-8 所示是 OpenFeign 发送网络请求时几个关键类的交互流程,可分为 3 个阶段:第 1 阶段是将函数实际参数值添加到 RequestTemplate 中;第 2 阶段是调用 Target 生成具体的 Request 对象;第 3 阶段是调用 Client 来发送网络请求,然后将 Response 转化为对象返回。

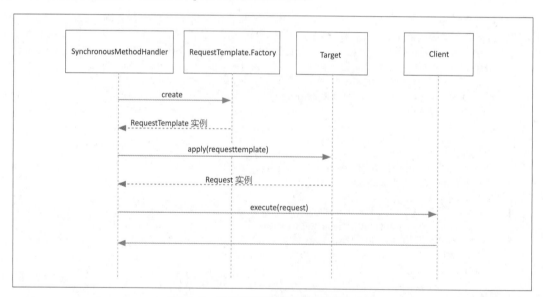

图6-8 OpenFeign发送网络请求时几个关键类的交互流程

invoke 方法的代码如下所示。

```
// SynchronousMethodHandler.java
    public Object invoke(Object[] argv) throws Throwable {
        RequestTemplate template = this.buildTemplateFromArgs.create(argv);
        Retryer retryer = this.retryer.clone();
```

```
      while(true) {
        try {
          return this.executeAndDecode(template);
        } catch (RetryableException var8) {
          RetryableException e = var8;

          try {
            retryer.continueOrPropagate(e);
          } catch (RetryableException var7) {
            Throwable cause = var7.getCause();
            if (this.propagationPolicy == ExceptionPropagationPolicy.UNWRAP && cause != null) {
              throw cause;
            }

            throw var7;
          }

          if (this.logLevel != Level.NONE) {
            this.logger.logRetry(this.metadata.configKey(), this.logLevel);
          }
        }
      }
    }
```

如上述代码所示，SynchronousMethodHandler 的 invoke 方法先创建了 RequestTemplate 对象。在该对象的创建过程中，使用到之前收集的函数信息 methodmetadata。遍历 methodmetadata 中与参数相关的 indexToName，然后根据参数的索引从 invoke 的参数数组中获得对应的值，将其填入对应的键值对中。最后依次处理和头部相关的参数值。invoke 方法调用 RequestTemplate.Factory 的 create 方法来创建 RequestTemplate 对象，代码如下所示。

```
    // RequestTemplate.Factory
public RequestTemplate create(Object[] argv) {
    RequestTemplate mutable = RequestTemplate.from(this.metadata.template());

      // 设置 URL
if (this.metadata.urlIndex() != null) {
        int urlIndex = this.metadata.urlIndex();
            Util.checkArgument(argv[urlIndex] != null, "URI parameter %s was null", new Object[]{urlIndex});
        mutable.target(String.valueOf(argv[urlIndex]));
    }

    Map<String, Object> varBuilder = new LinkedHashMap();
// 遍历 metadata 中所有关于参数的索引及其对应名称的配置信息
    Iterator var4 = this.metadata.indexToName().entrySet().iterator();

    while(true) {
      Map.Entry entry;
      int i;
      Object value;
      do {
        if (!var4.hasNext()) {
```

```java
            RequestTemplate template = this.resolve(argv, mutable, varBuilder);
            if (this.metadata.queryMapIndex() != null) {
                Object value = argv[this.metadata.queryMapIndex()];
                Map<String, Object> queryMap = this.toQueryMap(value);
                template = this.addQueryMapQueryParameters(queryMap, template);
            }

            if (this.metadata.headerMapIndex() != null) {
                template = this.addHeaderMapHeaders((Map)argv[this.metadata.headerMapIndex()], template);
            }

            return template;
        }

// 获取参数
            entry = (Map.Entry)var4.next();
// 参数的索引
            i = (Integer)entry.getKey();
// 参数的值
            value = argv[i];
        } while(value == null);

        if (this.indexToExpander.containsKey(i)) {
// 将值转换成字符串
            value = this.expandElements((Param.Expander)this.indexToExpander.get(i), value);
        }

        Iterator var8 = ((Collection)entry.getValue()).iterator();

        while(var8.hasNext()) {
            String name = (String)var8.next();
            varBuilder.put(name, value);
        }
    }
}
```

上述代码中的 resolve 方法首先会替换 URL 中的 pathValues，然后对 URL 进行编码，接着将所有头部信息进行转化，最后处理请求的 body 数据，如下所示。

```java
// RequestTemplate.Factory
public RequestTemplate resolve(Map<String, ?> variables) {
    StringBuilder uri = new StringBuilder();
    RequestTemplate resolved = from(this);
    if (this.uriTemplate == null) {
        this.uriTemplate = UriTemplate.create("", !this.decodeSlash, this.charset);
    }

    uri.append(this.uriTemplate.expand(variables));
    String headerValues;
    String queryString;
    if (!this.queries.isEmpty()) {
```

```java
      resolved.queries(Collections.emptyMap());
      StringBuilder query = new StringBuilder();
      Iterator queryTemplates = this.queries.values().iterator();

      while(queryTemplates.hasNext()) {
        QueryTemplate queryTemplate = (QueryTemplate)queryTemplates.next();
        headerValues = queryTemplate.expand(variables);
        if (Util.isNotBlank(headerValues)) {
          query.append(queryTemplate.expand(variables));
          if (queryTemplates.hasNext()) {
            query.append("&");
          }
        }
      }

      queryString = query.toString();
      if (!queryString.isEmpty()) {
        Matcher queryMatcher = QUERY_STRING_PATTERN.matcher(uri);
        if (queryMatcher.find()) {
          uri.append("&");
        } else {
          uri.append("?");
        }

        uri.append(queryString);
      }
    }

    resolved.uri(uri.toString());
    if (!this.headers.isEmpty()) {
      resolved.headers(Collections.emptyMap());
      Iterator var8 = this.headers.values().iterator();

      while(var8.hasNext()) {
        HeaderTemplate headerTemplate = (HeaderTemplate)var8.next();
        queryString = headerTemplate.expand(variables);
        if (!queryString.isEmpty()) {
          headerValues = queryString.substring(queryString.indexOf(" ") + 1);
          if (!headerValues.isEmpty()) {
            resolved.header(headerTemplate.getName(), headerValues);
          }
        }
      }
    }

    resolved.body(this.body.expand(variables));
    resolved.resolved = true;
    return resolved;
}
```

executeAndDecode 方法会根据 RequestTemplate 生成 Request 对象，然后交给 Client 实例发送网络请求，最后返回对应的函数返回类型的实例。executeAndDecode 方法的具体实现如下所示。

```java
// SynchronousMethodHandler.java
  Object executeAndDecode(RequestTemplate template) throws Throwable {
```

```java
// 根据 RequestTemplate 生成 Request 对象
    Request request = this.targetRequest(template);
    if (this.logLevel != Level.NONE) {
        this.logger.logRequest(this.metadata.configKey(), this.logLevel, request);
    }

    long start = System.nanoTime();

    Response response;

// client 发送网络请求,client 可能为 OkHttpClient 或 ApacheClient
    try {
        response = this.client.execute(request, this.options);
    } catch (IOException var15) {
        if (this.logLevel != Level.NONE) {
            this.logger.logIOException(this.metadata.configKey(), this.logLevel, var15, this.
                    elapsedTime(start));
        }

        throw FeignException.errorExecuting(request, var15);
    }

    long elapsedTime = TimeUnit.NANOSECONDS.toMillis(System.nanoTime() - start);
    boolean shouldClose = true;

    try {
        if (this.logLevel != Level.NONE) {
            response = this.logger.logAndRebufferResponse(this.metadata.configKey(), this.logLevel,
                    response, elapsedTime);
        }

// 如果 response 的类型就是函数的返回类型,那么可以直接返回
        if (Response.class == this.metadata.returnType()) {
            Response var18;
            if (response.body() == null) {
                var18 = response;
                return var18;
            } else if (response.body().length() != null && (long)response.body().length() <= 8192L) {

// 设置 body
                byte[] bodyData = Util.toByteArray(response.body().asInputStream());
                Response var20 = response.toBuilder().body(bodyData).build();
                return var20;
            } else {
                shouldClose = false;
                var18 = response;
                return var18;
            }
        } else {
            Object result;
            Object var10;
            if (response.status() >= 200 && response.status() < 300) {
                if (Void.TYPE == this.metadata.returnType()) {
                    result = null;
```

```java
          return result;
        } else {
          result = this.decode(response);
          shouldClose = this.closeAfterDecode;
          var10 = result;
          return var10;
        }
      } else if (this.decode404 && response.status() == 404 && Void.TYPE != this.metadata.returnType()) {
        result = this.decode(response);
        shouldClose = this.closeAfterDecode;
        var10 = result;
        return var10;
      } else {
        throw this.errorDecoder.decode(this.metadata.configKey(), response);
      }
    }
  } catch (IOException var16) {
    if (this.logLevel != Level.NONE) {
      this.logger.logIOException(this.metadata.configKey(), this.logLevel, var16, elapsedTime);
    }

    throw FeignException.errorReading(request, response, var16);
  } finally {
    if (shouldClose) {
      Util.ensureClosed(response.body());
    }

  }
}
```

OpenFeign 提供了 RequestInterceptor 机制,在由 RequestTemplate 生成 Request 对象的过程中,会调用所有 RequestInterceptor 对 RequestTemplate 进行处理。而 Target 是生成 JAX-RS 2.0 网络请求 Request 的接口类。RequestInterceptor 处理的具体实现如下所示。

```java
// SynchronousMethodHandler.java
Request targetRequest(RequestTemplate template) {

// 使用请求拦截器为每个请求添加固定的 Header 信息
    Iterator var2 = this.requestInterceptors.iterator();

    while(var2.hasNext()) {
      RequestInterceptor interceptor = (RequestInterceptor)var2.next();
      interceptor.apply(template);
    }

    return this.target.apply(template);
}
```

Client 是用来发送网络请求的接口类,有 OkHttpClient 和 RibbonClient 两个子类。OkHttpClient 调用 OkHttp 的相关组件进行网络请求的发送。OkHttpClient 的具体实现如下所示。

```java
// OkHttpClient.java
```

```
public Response execute(feign.Request input, Options options) throws IOException {
    okhttp3.OkHttpClient requestScoped;
    if (this.delegate.connectTimeoutMillis() == options.connectTimeoutMillis() && this.delegate.readTimeoutMillis() == options.readTimeoutMillis()) {
        requestScoped = this.delegate;
    } else {
        requestScoped = this.delegate.newBuilder().connectTimeout((long)options.connectTimeoutMillis(), TimeUnit.MILLISECONDS).readTimeout((long)options.readTimeoutMillis(), TimeUnit.MILLISECONDS).followRedirects(options.isFollowRedirects()).build();
    }

// 将 feign.Request 转换为 OkHttp 的 Request 对象
    Request request = toOkHttpRequest(input);

// 使用 OkHttp 的同步操作发送网络请求
    okhttp3.Response response = requestScoped.newCall(request).execute();

// 将 OkHttp 的 Response 转换成 Feign.Response
    return toFeignResponse(response, input).toBuilder().request(input).build();
}
```

第 7 章

Hystrix 服务熔断

在微服务架构中存在多个可直接调用的服务，这些服务若在调用时出现故障会导致连锁效应，也就是可能会让整个系统变得不可用，这种情况被称为服务雪崩效应。如何避免服务雪崩效应呢？这个问题可以使用 Hystrix 服务熔断来解决。

本章的主要内容如下。

1. 认识 Hystrix。
2. 使用 REST 实现服务熔断。
3. 使用 Feign 实现服务熔断。
4. 使用 Hystrix 实现监控。
5. 源码分析。

7.1 认识 Hystrix

Hystrix 是 Netflix 的一个开源项目，它能够在依赖服务失效的情况下，通过隔离系统依赖服务的方式，防止服务级联失败；同时 Hystrix 提供失败回滚机制，使系统能够更快地从异常中恢复。

Hystrix 官方简介如图 7-1 所示。

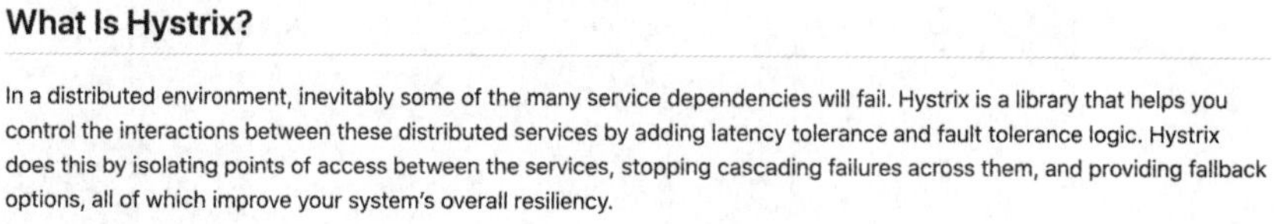

What Is Hystrix?

In a distributed environment, inevitably some of the many service dependencies will fail. Hystrix is a library that helps you control the interactions between these distributed services by adding latency tolerance and fault tolerance logic. Hystrix does this by isolating points of access between the services, stopping cascading failures across them, and providing fallback options, all of which improve your system's overall resiliency.

图7-1　Hystrix官方简介

图 7-1 中的简介翻译如下。

在分布式环境中，许多依赖服务中的一些不可避免地会失败。Hystrix 通过添加延迟容错和容错逻辑来帮助使用者控制这些分布式服务之间的交互。Hystrix 通过隔离服务之间的访问点、避免它们之间的级联故障并提供回退选项来提高系统的整体弹性。

7.1.1　雪崩效应

在微服务中，服务间的调用关系错综复杂，一个请求可能需要调用多个微服务接口才能实现，这会形成非常复杂的调用链路。正常服务调用链路如图 7-2 所示。

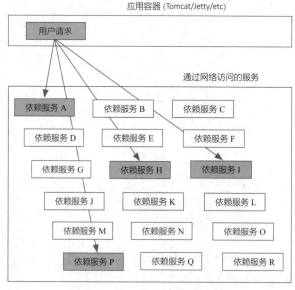

图7-2　正常服务调用链路

在图 7-2 中，一个业务请求，需要调用 A、P、H、I 4 个服务，这 4 个服务又可能调用其他服务。如果此时某个服务，比如 I 服务，发生异常，请求阻塞，用户请求就不会得到响应，则 Tomcat 服务器的这个线程不会释放。由于越来越多的用户请求到来，越来越多的线程会阻塞，如图 7-3 所示。

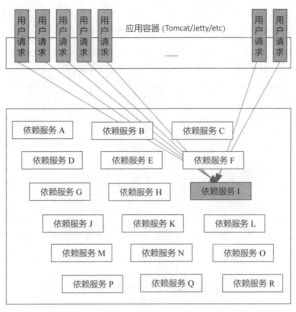

图7-3　异常服务调用链路

服务器支持的线程和并发数有限，请求一直阻塞，会使服务器资源耗尽，从而导致其他服务都不可用，出现雪崩效应。

这就好比，一条汽车生产线，生产不同的汽车需要使用不同的零件，如果某个零件因为种种原因无法使用，就会造成整辆车无法装配，陷入等待零件的状态，直到零件到位，才能继续组装。此时，如果有很多个车型都需要这个零件，那么整个工厂都将陷入等待的状态，导致所有生产线都瘫痪，一个零件的波及范围不断扩大。

虽然从源头上无法完全杜绝雪崩的发生，但是雪崩发生的根本原因在于服务之间的强依赖，因此开发者可以提前评估，做好熔断、隔离、限流。

Hystrix 解决雪崩问题的手段主要是服务降级，主要包括以下两种。

（1）线程隔离。

（2）服务熔断。

7.1.2　线程隔离

如图 7-4 所示，Hystrix 为每个依赖服务调用分配一个小的线程池，如果线程池已满，调用将被立即拒绝。默认不采用排队，使失败判定时间更短。

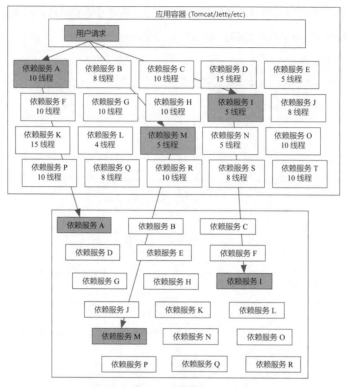

图7-4 线程隔离

用户请求将不再直接访问服务，而是通过线程池中的空闲线程来访问服务。如果线程池已满，或者请求超时，则会进行降级处理。服务降级要优先保证核心服务，而非核心服务不可用或弱可用。

7.1.3 服务熔断

熔断这一概念源于电子工程中的断路器。在互联网系统中，当下游服务因访问压力过大而响应变慢或失败时，上游服务为了维持系统整体的可用性，可以暂时切断对下游服务的调用。这种牺牲局部、保全整体的措施就叫作熔断，如图 7-5 所示。

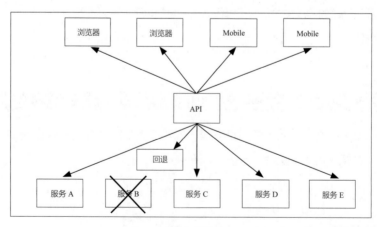

图7-5 熔断

在分布式架构中，断路器的作用也是类似的，当某个服务单元发生故障（类似用电器发生短路）之后，通过断路器的故障监控（类似熔丝），向调用方返回一个错误响应，而不是陷入长时间的等待。这样就不会使线程因调用故障服务被长时间占用，避免了故障在分布式系统中的蔓延。

用户请求发生故障时，不会被阻塞，更不会无休止地等待或者任由系统崩溃，至少可以有一个执行结果（例如返回友好的提示信息）。

服务降级虽然会导致请求失败，但是不会导致阻塞，而且最多只会影响这个依赖服务对应的线程池中的资源，对其他服务没有影响。

限流可以认为是服务降级的一种，就是通过限制系统的输入和输出流量以达到保护系统的目的。一般来说，系统的吞吐量是可以被测算的，为了保证系统的稳定运行，一旦吞吐量达到需要限制的阈值，就需要采取少量措施实现限流，比如推迟解决、拒绝解决，或者部分拒绝解决等。

触发 Hystrix 服务降级的情况如下。

（1）线程池已满。

（2）请求超时。

7.2 使用 REST 实现服务熔断

以 Eureka 为注册中心，订单微服务使用 REST 调用商品微服务。可启动两个商品微服务来实现负载均衡，端口号分别设置为 8001 和 8002。

7.2.1 坐标依赖

在服务消费者 order-service 中添加 Hystrix 依赖，如下所示。

```
<dependency>
    <groupId>org.springframework.cloud</groupId>
    <artifactId>spring-cloud-starter-netflix-hystrix</artifactId>
</dependency>
```

7.2.2 工程改造

将贯穿案例的部分代码进行修改，目前的项目结构如图 7-6 所示。

图7-6 项目结构

■ 开启断路器

修改启动类 order-service#OrderApplication，代码如下。

```
@SpringBootApplication
@EntityScan("cn.book.order.entity")
@EnableCircuitBreaker // 开启断路器
public class OrderApplication {
  @Bean
  @LoadBalanced
  public RestTemplate restTemplate(){
     return new RestTemplate();
  }


  public static void main(String[] args) {
     SpringApplication.run(OrderApplication.class, args);
  }
}
```

在启动类 OrderApplication 头部添加 @EnableCircuitBreaker 注解开启断路器。可以看到，目前启动类上的注解越来越多。在微服务中，经常会引入 @SpringBootApplication、@EnableDiscoveryClient 和 @EnableCircuitBreaker 这 3 个注解，于是 Spring 提供了一个组合注解 @SpringCloudApplication，源码如下。

```
@Target({ElementType.TYPE})
@Retention(RetentionPolicy.RUNTIME)
@Documented
@Inherited
@SpringBootApplication
@EnableDiscoveryClient
@EnableCircuitBreaker
public @interface SpringCloudApplication {
}
```

■ 配置熔断降级业务逻辑

修改 order-service#OrderController，添加 fallbackMethod 属性，在 order 方法头部添加 @HystrixCommand 并指定 fallbackMethod 属性。

```
@RequestMapping(value = "/{product_id}/{product_num}", method = RequestMethod.GET)
@HystrixCommand(fallbackMethod = "orderFallBack")
public Map<String, Object> order(@PathVariable Long product_id, @PathVariable Integer product_num) {
  // 定义 URL
  String url = String.format("http://%s/product/%d", "product-service", product_id);
  // 使用 restTemplate 调用 HTTP 接口
  Product product = restTemplate.getForEntity(url, Product.class).getBody();
  // 创建字典对象，存放信息
  Map<String, Object> map = new HashMap<String, Object>();
  // 存数据
  map.put("product_num", product_num);
  map.put("product", product);
  // 返回数据
```

```
        return map;
    }
    public Map<String,Object> orderFallBack(Long product_id, Integer product_num) {
        Product product = new Product();
        product.setId(-1l);
        product.setProductName(" 熔断：触发降级方法 ");

        Map<String, Object> map = new HashMap<String, Object>();
        map.put("product_num", product_num);
        map.put("product", product);
        return map;
    }
```

由上述代码可知，为 order 方法编写一个回退方法 orderFallBack，该方法与 order 方法具有相同的参数与返回值类型，返回一个默认的错误信息。这里的参数和返回类型如果不一致，会出现图 7-7 所示的异常信息。

Whitelabel Error Page

This application has no explicit mapping for /error, so you are seeing this as a fallback.

Tue Apr 06 17:33:32 IRKT 2021
There was an unexpected error (type=Internal Server Error, status=500).
Incompatible return types. Command method: public java.util.Map cn.book.order.controller.OrderController.order(java.lang.Long,java.lang.Integer); Fallback method: public cn.book.order.entity.Product cn.book.order.controller.OrderController.orderFallBack(java.lang.Long,java.lang.Integer); Hint: Different size of types variables. Command type literals size = 3: [java.util.Map<java.lang.String, java.lang.Object>, class java.lang.String, class java.lang.Object] Fallback type literals size = 1: [class cn.book.order.entity.Product] Command type literal pos: unknown; Fallback type literal pos: unknown

图7-7 异常信息

在 order 方法上，使用注解 @HystrixCommand 的 fallbackMethod 属性，指定熔断触发的降级方法是 orderFallBack。

■ 模拟超时

修改 product-service-8001#ProductController#findById 方法，如下所示。

```
@RequestMapping(value = "/{id}", method = RequestMethod.GET)
public Product findById(@PathVariable Long id) throws InterruptedException {
    // 模拟超时时间
    Thread.sleep(2000);
    Product product = productService.findById(id);
    product.setProductName(" 访问的服务地址信息： " + ip + ":" + port);
    return product;
}
```

Hystrix 的默认超时时间为 1s，可以通过配置修改这个值。默认超时时间配置如下所示。

```
hystrix:
  command:
    default:
      execution:
        isolation:
```

```
      thread:
        timeoutInMilliseconds: 1000
```

7.2.3 代码测试

重启各个服务后，访问订单微服务，结果如图 7-8 所示。

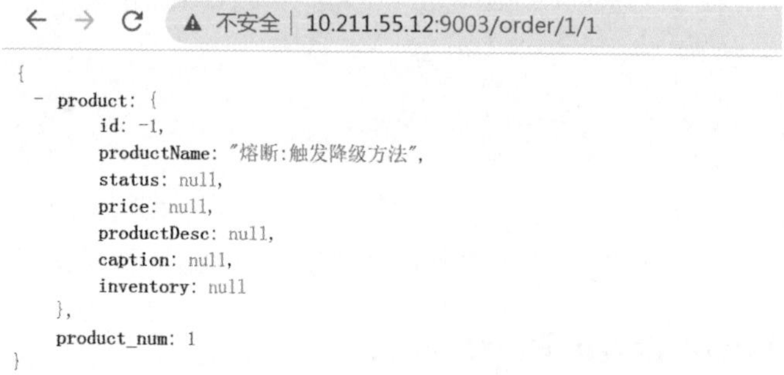
不安全 | 10.211.55.12:9003/order/1/1
{
- product: {
id: -1,
productName: "熔断:触发降级方法",
status: null,
price: null,
productDesc: null,
caption: null,
inventory: null
},
product_num: 1
}

图7-8 访问订单微服务——熔断降级

再次访问订单微服务，结果如图 7-9 所示。

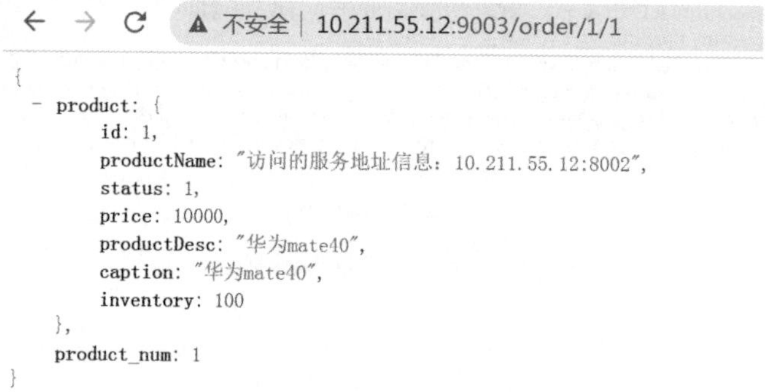
不安全 | 10.211.55.12:9003/order/1/1
{
- product: {
id: 1,
productName: "访问的服务地址信息：10.211.55.12:8002",
status: 1,
price: 10000,
productDesc: "华为mate40",
caption: "华为mate40",
inventory: 100
},
product_num: 1
}

图7-9 访问订单微服务——正常

负载均衡默认采用轮询的方式，订单微服务调用了两次商品微服务。第一次调用的是端口号为 8001 的商品微服务，内部有 2s 延迟，大于 Hystrix 的默认超时时间 1s，所以此时订单微服务启动熔断降级，调用 orderFallBack 方法并将其访问值作为响应返回。第二次调用的是端口号为 8002 的商品微服务，正常响应。

7.3 使用 Feign 实现服务熔断

以 Eureka 为注册中心，订单微服务使用 Feign 调用商品微服务。可启动两个商品微服务来完成负载均衡，端口号分别设置为 8001 和 8002。

7.3.1 坐标依赖

Spring Cloud Feign 默认已为 Feign 整合了 Hystrix，所以添加 Feign 依赖后就不用再添加 Hystrix 依赖。如图 7-10 所示，org.springframework.cloud:spring-cloud-starter-openfeign:2.1.0.RELEASE 包中已经包含了 io.github.openfeign:feign-hystrix:10.1.0，所以这里不需要导入任何额外的坐标。

```
> eureka-server
∨ order-service
    > Lifecycle
    > Plugins
    ∨ Dependencies
        mysql:mysql-connector-java:5.1.32
      > org.springframework.boot:spring-boot-starter-data-jpa:2.1.6.RELEASE
      > org.springframework.cloud:spring-cloud-starter-netflix-eureka-client:2.1.0.RELEASE
      ∨ org.springframework.cloud:spring-cloud-starter-openfeign:2.1.0.RELEASE
            org.springframework.cloud:spring-cloud-starter:2.1.0.RELEASE (omitted for dupli
         > org.springframework.cloud:spring-cloud-openfeign-core:2.1.0.RELEASE
         > org.springframework:spring-web:5.1.8.RELEASE
         > org.springframework.cloud:spring-cloud-commons:2.1.0.RELEASE
            io.github.openfeign:feign-core:10.1.0
         > io.github.openfeign:feign-slf4j:10.1.0
         > io.github.openfeign:feign-hystrix:10.1.0
      > io.github.openfeign:feign-okhttp:10.1.0
      > org.springframework.boot:spring-boot-starter-web:2.1.6.RELEASE
      > org.springframework.boot:spring-boot-starter-logging:2.1.6.RELEASE
      > org.springframework.boot:spring-boot-starter-test:2.1.6.RELEASE (test)
         org.projectlombok:lombok:1.18.4 (provided)
> product-service-8001
> product-service-8002 (root)
> shop-parent (root)
```

图7-10　坐标依赖

7.3.2 工程改造

将贯穿案例的部分代码进行修改，目前的项目结构如图 7-11 所示。

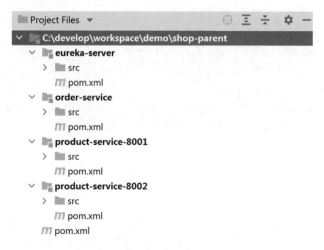

图7-11　项目结构

■ 开启断路器

在 Feign 中已经内置了 Hystrix，但是 Hystrix 默认是关闭的，需要在工程的 application.yml 中开启对 Hystrix 的支持。修改 order-service#application.yml，如下所示。

```yaml
feign:
  hystrix: # 在 Feign 中开启 Hystrix
    enabled: true
```

■ 创建 FeignClient 接口的实现类

基于 Feign 实现熔断降级，那么降级方法需要创建到 FeignClient 接口的实现类中，在实现类头部加上注解 @Component，如下所示。

```java
/**
 * 实现自定义的 ProductFeignClient 接口
 * 在接口实现类中编写熔断降级方法
 */
@Component
public class ProductFeignClientFallBack implements ProductFeignClient {
    @Override
    /**
     * 降级方法
     */
    public Product findById(Long id) {
        Product product = new Product();
        product.setId(-1L);
        product.setProductName(" 熔断：触发降级方法 ");
        return product;
    }
}
```

■ 修改 FeignClient 添加 Hystrix 熔断

在接口 Product FeignClient 头部添加注解 @FeignClient，并在注解中添加降级方法，如下所示。

```java
@FeignClient(
    value = "product-service",// 服务提供者名称
    configuration = FeignConfiguration.class,// 配置类
    fallback = ProductFeignClientFallback.class// 熔断降级方法实现类
)
public interface ProductFeignClient {
    @RequestMapping(value = "/product/{id}", method = RequestMethod.GET)
    public Product findById(@PathVariable Long id);
}
```

■ 模拟超时

修改 product-service-8001#ProductController#findById 方法，如下所示。

```java
@RequestMapping(value = "/{id}", method = RequestMethod.GET)
public Product findById(@PathVariable Long id) throws InterruptedException {
    // 模拟超时时间
```

```
Thread.sleep(2000);
Product product = productService.findById(id);
product.setProductName(" 访问的服务地址信息: " + ip + ":" + port);
return product;
}
```

7.3.3 代码测试

重启各个服务后，访问订单微服务，结果如图 7-12 所示。

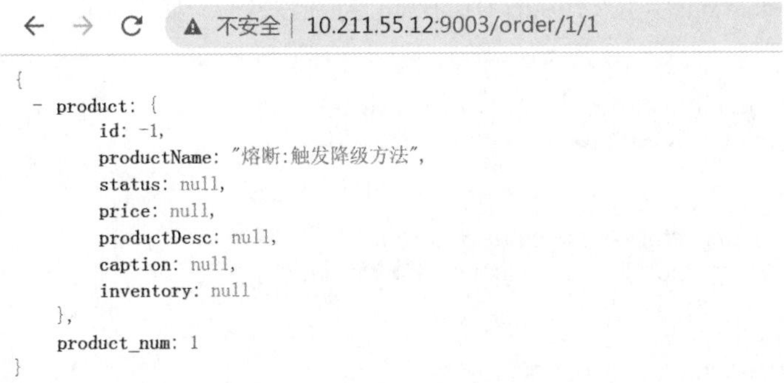

图7-12 访问订单微服务——熔断降级

再次访问订单微服务，结果如图 7-13 所示。

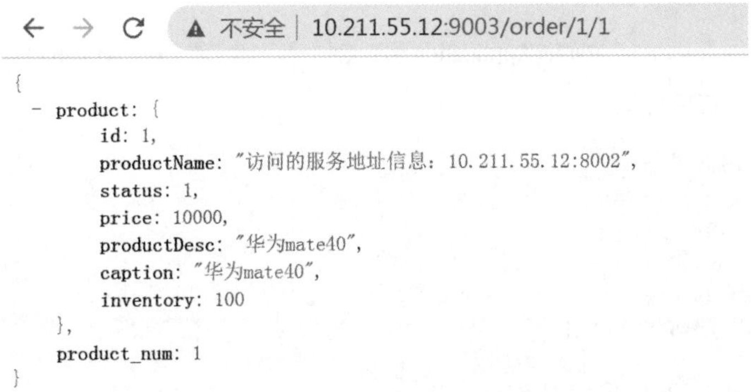

图7-13 访问订单微服务——正常

负载均衡默认采用轮询的方式，订单微服务调用了两次商品微服务。第一次调用的是端口号为 8001 的商品微服务，内部有 2s 延迟，大于 Hystrix 的默认超时时间 1s，所以此时订单微服务启动熔断降级，调用 FallBack 实现方法并将其访问值作为响应返回。第二次调用的是端口号为 8002 的商品微服务，正常响应。

7.4 使用 Hystrix 实现监控

在微服务架构中，当请求失败、被拒绝或超时的时候都会进入降级方法。但进入降级方法并不意味着断路器已经被打开。可以通过 Hystrix 相关的监控平台了解断路器的状态。

7.4.1 使用 Hystrix Dashboard 查看监控数据

Hystrix 除了实现容错,还提供实时监控功能。在调用服务时,Hystrix 会实时记录关于 Hystrix Command 的执行信息,比如每秒的请求数、成功数等,可以通过监控平台直观地查看实时的监控数据。

■ 引入坐标

在 order-service#pom.xml 中引入 Hystrix Dashboard 监控相关坐标,如下所示。

```xml
<dependency>
    <groupId>org.springframework.boot</groupId>
    <artifactId>spring-boot-starter-actuator</artifactId>
</dependency>
<dependency>
    <groupId>org.springframework.cloud</groupId>
    <artifactId>spring-cloud-starter-netflix-hystrix</artifactId>
</dependency>
<dependency>
    <groupId>org.springframework.cloud</groupId>
    <artifactId>spring-cloud-starter-netflix-hystrix-dashboard</artifactId>
</dependency>
```

■ 添加注解

在启动类 order-service#OrderApplication 头部添加注解 @EnableHystrixDashboard 和 @EnableCircuitBreaker,如下所示。

```java
@SpringBootApplication
@EntityScan("cn.book.order.entity")
@EnableFeignClients
@EnableHystrixDashboard
@EnableCircuitBreaker
public class OrderApplication {

    public static void main(String[] args) {
        SpringApplication.run(OrderApplication.class, args);
    }
}
```

■ 访问测试

重启各个服务后,首次访问 actuator,如图 7-14 所示,目前是没有数据的。

图7-14　首次访问actuator

多次访问 /order/1/1 后，再次访问 actuator，如图 7-15 所示，可以看到监控的数据。

图7-15　再次访问actuator

访问 actuator 获取的都是以文字形式展示的信息。很难通过文字直观地展示系统的运行状态，所以 Hystrix 官方还提供了图形化的 Dashboard 监控平台。Hystrix Dashboard 可以显示每个断路器（被 @HystrixCommand 注解的方法）的状态，对监控进行图形化展示。

访问 Hystrix Dashboard 首页，如图 7-16 所示。

图7-16　Hystrix Dashboard首页

在图 7-16 所示页面最上面的文本框中输入 actuator 的地址,如图 7-17 所示。

图7-17　输入actuator的地址

在图 7-17 所示的页面中单击【Monitor Stream】按钮,跳转到 Hystrix Dashboard 监控数据展示页面,如图 7-18 所示。

图7-18　Hystrix Dashboard监控数据展示页面

多次访问 /order/1/1 后,再次访问 Hystrix Dashboard 监控数据展示页面,如图 7-19 所示,可以看到监控的数据。

图7-19　再次访问Hystrix Dashboard监控数据展示页面

图 7-19 中的数字"1、0、0、2、0、0、66.0%"与图中右侧单词"Success、Short-Circuited、Bad Request、Timeout、Rejected、Failure、Error %"一一对应,比如 Success 的次数是 1,Timeout 的次数是 2 等。

7.4.2 使用 Hystrix Turbine 聚合监控数据

在微服务架构中,每个服务都需要配置 Hystrix Dashboard 监控。如果每次只能查看单个实例的监控数据,就需要不断切换监控地址,这显然很不方便。要想看整个系统的 Hystrix Dashboard 监控数据就需要用到 Hystrix Turbine。Hystrix Turbine 是一个聚合 Hystrix Dashboard 监控数据的工具,它可以将所有相关微服务的 Hystrix Dashboard 监控数据聚合到一起,以方便使用和查看。引入 Hystrix Turbine 后,整个监控系统架构如图 7-20 所示。

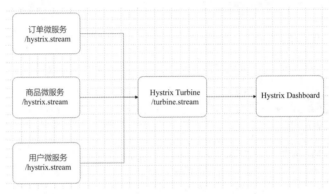

图7-20 整个监控系统架构

■ 创建工程

使用 IntelliJ IDEA 创建子工程——Turbine 微服务(turbine-service),如图 7-21 所示。

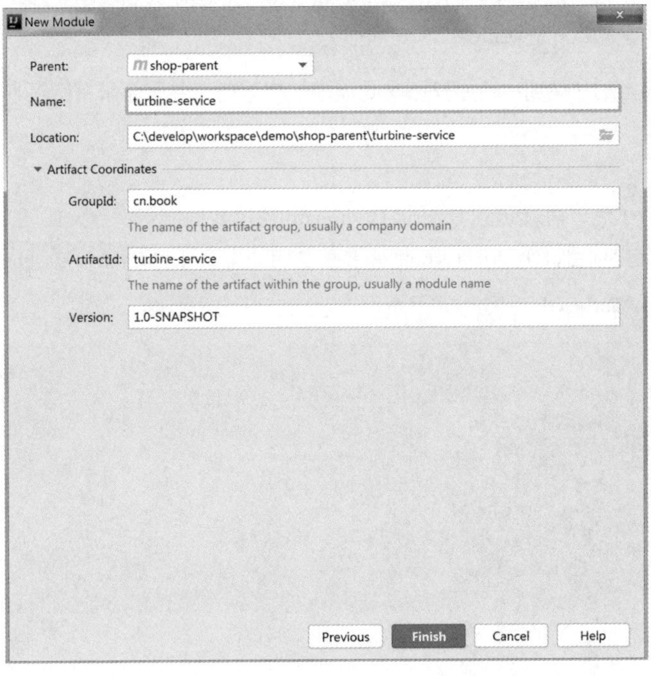

图7-21 Turbine微服务

■ 引入坐标

在 turbine-service#pom.xml 中引入 Hystrix Turbine 监控相关坐标，如下所示。

```xml
<dependency>
    <groupId>org.springframework.cloud</groupId>
    <artifactId>spring-cloud-starter-netflix-turbine</artifactId>
</dependency>
<dependency>
    <groupId>org.springframework.cloud</groupId>
    <artifactId>spring-cloud-starter-netflix-hystrix</artifactId>
</dependency>
<dependency>
    <groupId>org.springframework.cloud</groupId>
    <artifactId>spring-cloud-starter-netflix-hystrix-dashboard</artifactId>
</dependency>
```

■ 配置多个微服务的 Hystrix Turbine 监控

在 turbine-service#application.yml 配置文件中开启 Hystrix Turbine 并进行相关配置，如下所示。

```yaml
server:
  port: 9004 # 端口
spring:
  application:
    name: turbine-service # 服务名称
eureka: # 配置 Eureka
  client:
    service-url:
      defaultZone: http://10.211.55.12:9000/eureka/ # 多个 Eureka 服务端之间用逗号隔开
  instance:
    prefer-ip-address: true # 使用 IP 地址注册
    instance-id: ${spring.cloud.client.ip-address}:${server.port} # 向注册中心注册服务 id
turbine:
  app-config: order-service # 要监控的微服务列表，多个微服务之间使用逗号分隔
  cluster-name-expression: "'default'"
```

Hystrix Turbine 会自动从注册中心中获取需要监控的微服务，并聚合所有微服务中的 /hystrix.stream 数据。

■ 配置启动类

作为一个独立的监控项目，Hystrix Turbine 需要配置启动类，开启 Hystrix Dashboard 监控平台，并激活 Hystrix Turbine。Hystrix Turbine 的启动类如下所示。

```java
@SpringBootApplication
@EnableTurbine
@EnableHystrixDashboard
public class TurbineApplication {

    public static void main(String[] args) {
        SpringApplication.run(TurbineApplication.class, args);
    }
}
```

■ 访问测试

重启各个服务后，访问 /hystrix，结果如图 7-22 所示。

图7-22　Hystrix Dashboard首页

在图 7-22 所示页面最上面的文本框中输入 /turbine.stream 的地址，如图 7-23 所示。

图7-23　输入地址

在图 7-23 所示的页面中单击【Monitor Stream】按钮，跳转到图形化展示页面，如图 7-24 所示。

图7-24　Hystrix Turbine监控数据展示

如果配置了监控多个微服务功能，图 7-24 中会显示对应的微服务的监控内容。

7.4.3　断路器的状态

断路器有 3 个状态：CLOSED（关闭）、OPEN（打开）和 HALF OPEN（半开）。断路器默认为关闭状态。当触发熔断后状态变更为打开状态，在等待到指定的时间后 Hystrix 会发送请求以检测服务是否开启，这期间断路器会变为半开状态，熔断探测服务可用则继续变更为关闭状态。3 种状态的转换如图 7-25 所示。

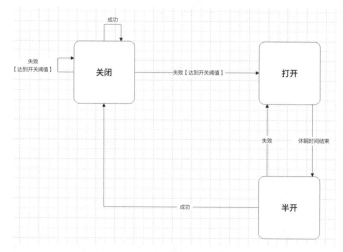

图 7-25　3 种状态的转换

CLOSED：断路器关闭状态，所有请求都能正常访问。代理类维护了最近调用失败的次数，如果某次调用失败，则使失败次数加 1。如果最近失败次数超过了在给定时间内允许失败次数的阈值，代理类则切换到断开（断路器打开）状态。此时代理类开启一个超时时钟，当该时钟超过了超时时间，则切换到半断开（断路器半开）状态。该超时时间的设定给了系统一次机会来修正导致调用失败的错误。

OPEN：断路器打开状态，所有请求都会被降级。Hystrix 会对请求进行计数，当一定时间内请求失败百分比达到阈值，则触发熔断，断路器会打开。默认失败百分比的阈值是 50%，请求次数不低于 20 次。

HALF OPEN：断路器半开状态，断路器打开状态不是永久的，打开后断路器会进入休眠时间（默认是 5s）。随后断路器会自动进入半开状态。此时断路器会释放 1 次请求通过，若这个请求是健康的，则会关闭断路器，否则继续保持打开，再次进行 5s 休眠计时。

下面通过一个案例测试断路器状态的变化。在商品微服务中加入业务逻辑，如下所示。

（1）传入的商品 id 非 0，则正常调用商品微服务。

（2）传入的商品 id 是 0，则抛出异常。

具体实现如下所示。

```
@RequestMapping(value = "/{id}", method = RequestMethod.GET)
public Product findById(@PathVariable Long id) {
    // 加入业务逻辑，测试断路器状态，如果商品 id 是 0 则抛出异常
    if (id == 0) {
        throw new RuntimeException(" 自定义抛出异常，测试断路器状态 ");
    }
    Product product = productService.findById(id);
    product.setProductName(" 访问的服务地址信息： "+ip + ":" + port);
    return product;
}
```

重启各个服务后，访问 /order/1/1，正常响应，如图 7-26 所示。

```
{
  - product: {
        id: 1,
        productName: "访问的服务地址信息: 10.211.55.12:8002",
        status: 1,
        price: 10000,
        productDesc: "华为mate40",
        caption: "华为mate40",
        inventory: 100
    },
    product_num: 1
}
```

图7-26　正常响应

访问 /order/0/1，发生异常导致熔断降级，如图 7-27 所示。

```
{
  - product: {
        id: -1,
        productName: "熔断:触发降级方法",
        status: null,
        price: null,
        productDesc: null,
        caption: null,
        inventory: null
    },
    product_num: 1
}
```

图7-27　熔断降级

访问 Hystrix Turbine，断路器关闭，如图 7-28 所示。

图7-28　访问Hystrix Turbine，断路器关闭

短时间内多次访问 /order/0/1，因为断路器打开的阈值是短时间内请求（请求次数不低于 20 次）成功率低于 50%。此时访问 Hystrix Turbine，断路器打开，如图 7-29 所示。

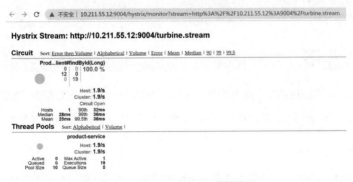

图7-29　访问Hystrix Turbine，断路器打开

再次访问 /order/1/1，结果如图 7-30 所示，由之前的正常响应变成现在的熔断降级。

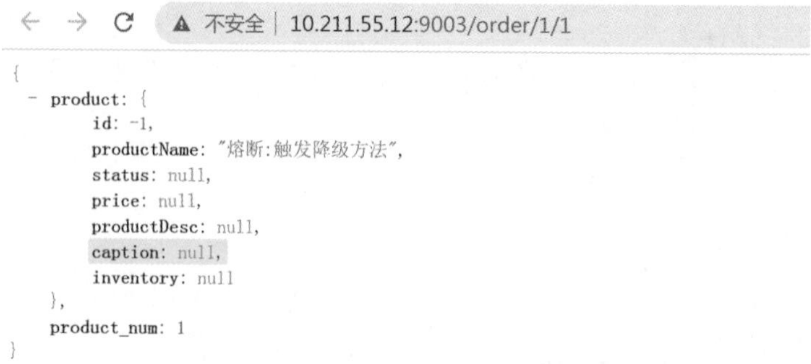
不安全 | 10.211.55.12:9003/order/1/1
{
- product: {
id: -1,
productName: "熔断:触发降级方法",
status: null,
price: null,
productDesc: null,
caption: null,
inventory: null
},
product_num: 1
}

图7-30　熔断降级

等待 5s 之后，再次访问 /order/1/1，结果如图 7-31 所示，又可以正常响应了。

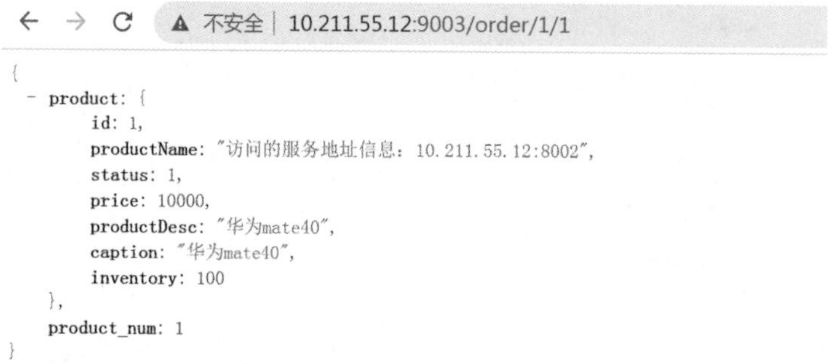
不安全 | 10.211.55.12:9003/order/1/1
{
- product: {
id: 1,
productName: "访问的服务地址信息：10.211.55.12:8002",
status: 1,
price: 10000,
productDesc: "华为mate40",
caption: "华为mate40",
inventory: 100
},
product_num: 1
}

图7-31　正常响应

默认触发熔断的最少请求次数是 20 次，休眠时间为 5s，可以通过配置修改熔断策略，如下所示。

```
hystrix:
 command:
  default:
   execution:
    isolation:
     thread:
      timeoutInMilliseconds: 2000 # 默认超时时间
   circuitBreaker:
    errorThresholdPercentage: 50 # 触发熔断错误的百分比阈值，默认值为 50%
    sleepWindowInMilliseconds: 10000 # 熔断后休眠时间，默认值为 5s
    requestVolumeThreshold: 10 # 触发熔断的最少请求次数，默认值为 20 次
```

7.4.4　断路器的隔离策略

微服务使用 Hystrix 断路器实现了服务的自动降级，使微服务具备了自我保护的能力，提升了系统的稳定性，也较好地解决了雪崩问题。其使用方式目前支持两种策略：线程池隔离和信号量隔离策略。

线程池隔离策略

线程池隔离策略是指使用一个线程池来存储当前的请求，通过线程池对请求进行处理，设置任务返回

处理超时时间，并将堆积的请求放入线程池队列。采用这种策略需要为每个依赖的服务申请线程池，有一定的资源消耗，好处是可以应对突发流量（流量洪峰来临时，处理不完的数据可存储到线程池队里慢慢处理）。

■ **信号量隔离策略**

信号量隔离策略是指使用一个原子计数器（或信号量）来记录当前有多少个线程在运行，请求来时先判断计数器的值，若超过设置的最大线程数则丢弃该类型的新请求，否则执行计数操作，即请求来时计数器加1，请求返回时计数器减1。采用这种策略需要严格地控制线程且立即返回模式，无法应对突发流量（流量洪峰来临时，处理的线程超过最大值，其他的请求会直接返回，不去请求依赖的服务）。

线程池隔离和信号量隔离两种策略的功能支持对比如表7-1所示。

表7-1 线程池隔离和信号量隔离两种策略的功能支持对比

隔离方式	是否支持超时	是否支持熔断	隔离原理	是否异步调用	资源消耗
线程池隔离	支持，可直接返回	支持，当线程池到达maxSize后，再请求会触发fallback接口进行熔断	每个服务单独用线程池	可以异步调用，也可以同步调用，看调用的方法	大，大量线程的上下文切换容易造成机器负载高
信号量隔离	不支持，如果阻塞，只能通过调用协议（如socket超时才能返回）	支持，当信号量达到maxconcurrentRequests后，再请求会触发fallback接口进行熔断	通过信号量的计数器来判断	同步调用，不支持异步	小，只是一个计数器

可以通过配置修改断路器的隔离策略为信号量策略，如下所示。

```
hystrix:
  command:
    default:
      execution:
        isolation:
          strategy: ExecutionIsolationStrategy.SEMAPHORE # 信号量策略
          maxConcurrentRequests: 100 # 最大信号量
          thread:
            timeoutInMilliseconds: 2000 # 默认超时时间
      circuitBreaker:
        errorThresholdPercentage: 50 # 触发熔断错误百分比阈值，默认值为50%
        sleepWindowInMilliseconds: 10000 # 熔断后休眠时间，默认值为5s
        requestVolumeThreshold: 10 # 触发熔断的最少请求次数，默认值为20次
```

7.5 源码分析

使用Hystrix后的远程调用流程如图7-32所示。简单的远程调用流程如下。

(1)构建 HystrixCommand 对象或者 HystrixObservableCommand 对象。

(2)执行方法。

(3)检查是否有相同命令执行的缓存。

(4)检查断路器是否打开。

(5)检查线程池或者信号量是否被消耗完。

(6)调用 HystrixObservableCommand#construct 或 HystrixCommand#run 执行被封装的远程调用方法。

(7)计算链路的健康情况。

(8)在命令执行失败时获取 Fallback 方法。

(9)返回成功的 Observable。

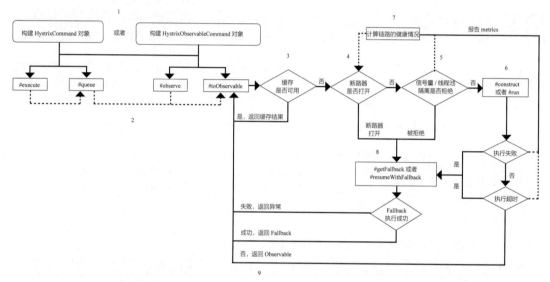

图7-32 远程调用流程

接着我们通过源码来逐步理解这些流程。

7.5.1 封装 HystrixCommand

@HystrixCommand 注解

在基础应用中使用 @HystrixCommand 注解来封装需要保护的远程调用方法。首先查看该注解的相关属性，代码如下所示。

```
//HystrixCommand.java

@Target({ElementType.METHOD})
@Retention(RetentionPolicy.RUNTIME)
@Inherited
@Documented
public @interface HystrixCommand {
    String groupKey() default "";
```

```java
    String commandKey() default "";
    String threadPoolKey() default "";
// 指定 Fallback 方法名，Fallback 方法也可以被 @HystrixCommand 注解
    String fallbackMethod() default "";
    HystrixProperty[] commandProperties() default {};
    HystrixProperty[] threadPoolProperties() default {};
    Class<? extends Throwable>[] ignoreExceptions() default {};
    ObservableExecutionMode observableExecutionMode() default ObservableExecutionMode.EAGER;
    HystrixException[] raiseHystrixExceptions() default {};
    String defaultFallback() default "";
}
```

一般来说，对于 HystrixCommand 的配置，仅需要关注 fallbackMethod 方法，当然如果对命令和线程池有特定需要，可以进行额外的配置。

除了 @HystrixCommand，还有一个 @HystrixCollapser 注解用于请求合并操作，但是需要与 @HystrixCommand 结合使用。合并操作的方法必须被 @HystrixCommand 注解，如下所示。

```java
//HystrixCollapser.java

@Target({ElementType.METHOD})
@Retention(RetentionPolicy.RUNTIME)
@Documented
public @interface HystrixCollapser {
    String collapserKey() default "";

    String batchMethod();

    Scope scope() default Scope.REQUEST;

    HystrixProperty[] collapserProperties() default {};
}
```

■ HystrixCommandAspect 切面

被注解修饰的方法将会被 HystrixCommand 封装调用，在 Hystrix 中通过 Aspect 切面的方式将被注解修饰的方法进行封装调用。具体代码如下所示。

```java
//HystrixCommandAspect.java

// 切面定义
@Around("hystrixCommandAnnotationPointcut() || hystrixCollapserAnnotationPointcut()")
public Object methodsAnnotatedWithHystrixCommand(ProceedingJoinPoint joinPoint) throws Throwable {
    Method method = AopUtils.getMethodFromTarget(joinPoint);
    Validate.notNull(method, "failed to get method from joinPoint: %s", new Object[]{joinPoint});
    if (method.isAnnotationPresent(HystrixCommand.class) && method.isAnnotationPresent(HystrixCollapser.class)) {
        throw new IllegalStateException("method can not be annotated with HystrixCommand and HystrixCollapser annotations at the same time");
    } else {
// 通过工厂的方式构建 metaHolderFactory
```

```java
        HystrixCommandAspect.MetaHolderFactory metaHolderFactory = (HystrixCommandAspect.
MetaHolderFactory)META_HOLDER_FACTORY_MAP.get(HystrixCommandAspect.HystrixPointcutType.
of(method));
        MetaHolder metaHolder = metaHolderFactory.create(joinPoint);
        HystrixInvokable invokable = HystrixCommandFactory.getInstance().create(metaHolder);
        ExecutionType executionType = metaHolder.isCollapserAnnotationPresent() ? metaHolder.
getCollapserExecutionType() : metaHolder.getExecutionType();

        try {
            Object result;
            if (!metaHolder.isObservable()) {
                result = CommandExecutor.execute(invokable, executionType, metaHolder);
            } else {
                result = this.executeObservable(invokable, executionType, metaHolder);
            }

            return result;
        } catch (HystrixBadRequestException var9) {
            throw var9.getCause();
        } catch (HystrixRuntimeException var10) {
            throw this.hystrixRuntimeExceptionToThrowable(metaHolder, var10);
        }
    }
}
```

上述代码主要执行步骤如下。

（1）通过 MetaHolderFactory 构建出被注解修饰的方法中用于构建 HystrixCommand 的必要信息集合类 MetaHolder。

（2）根据 MetaHolder，通过 HystrixCommandFactory 构建出合适的 HystrixCommand。

（3）委托 CommandExecutor 执行 HystrixCommand，得到结果。

MetaHolder 持有用于构建 HystrixCommand 和与被封装方法相关的必要信息，如被注解的方法、失败回滚执行的方法和默认的命令键等属性。其属性代码如下所示。

```java
//MetaHolder.java

@Immutable
public final class MetaHolder {
    ...
// 被注解的方法
    private final Method method;
    private final Method cacheKeyMethod;
    private final Method ajcMethod;
// 失败回滚执行的方法
    private final Method fallbackMethod;
    private final Object obj;
    private final Object proxyObj;
    private final Object[] args;
    private final Closure closure;
// 默认的 group 键
    private final String defaultGroupKey;
// 默认的命令键
```

```java
    private final String defaultCommandKey;
// 默认的合并请求键
    private final String defaultCollapserKey;
// 默认的线程池键
    private final String defaultThreadPoolKey;
// 执行类型
    private final ExecutionType executionType;
    ...
}
```

在 HystrixCommandFactory 类中，用于创建 HystrixCommand 的方法如下所示。

```java
// HystrixCommandFactory.java

public class HystrixCommandFactory {
    ...
    public HystrixInvokable create(MetaHolder metaHolder) {
        Object executable;
// 构建请求合并的命令
        if (metaHolder.isCollapserAnnotationPresent()) {
            executable = new CommandCollapser(metaHolder);
        } else if (metaHolder.isObservable()) {
            executable = new GenericObservableCommand(HystrixCommandBuilderFactory.getInstance().create(metaHolder));
        } else {
            executable = new GenericCommand(HystrixCommandBuilderFactory.getInstance().create(metaHolder));
        }

        return (HystrixInvokable) executable;
    }
}
```

根据 MetaHolder#isObservable 方法返回的不同属性，构建不同的命令，比如 HystrixCommand 或者 HystrixObservableCommand，前者将同步或者异步执行命令，后者异步回调执行命令。Hystrix 根据被封装方法的返回值类型来决定命令的执行方式，判断代码如下所示。

```java
// CommandMetaHolderFactory.java

...
ExecutionType executionType = ExecutionType.getExecutionType(method.getReturnType());
...

//ExecutionType.java

public enum ExecutionType {
// 异步执行命令
    ASYNCHRONOUS,
// 同步执行命令
    SYNCHRONOUS,
// 异步回调执行命令
    OBSERVABLE;

    private ExecutionType() {
```

```
    }
// 根据方法的返回值类型返回对应的 ExecutionType
    public static ExecutionType getExecutionType(Class<?> type) {
        if (Future.class.isAssignableFrom(type)) {
            return ASYNCHRONOUS;
        } else {
            return Observable.class.isAssignableFrom(type) ? OBSERVABLE : SYNCHRONOUS;
        }
    }
}
```

根据被封装方法的返回值类型决定命令执行的 ExecutionType，从而决定构建 HystrixCommand 还是 HystrixObservableCommand。其中 Future 类型的返回值会被异步执行，rx 类型的返回值会被异步回调执行，其他类型的返回值会被同步执行。

CommandExecutor 根据 MetaHolder 中 ExecutionType（执行类型）的不同，选择同步执行、异步执行还是异步回调执行，返回不同的执行结果。同步执行，直接返回结果对象；异步执行，返回 Future，封装了异步操作的结果；异步回调执行将返回 Observable，封装了响应式执行的结果，可以通过它对执行结果进行订阅，在执行结束后进行特定的操作。

图 7-33 所示为本节介绍的类的相关类图。

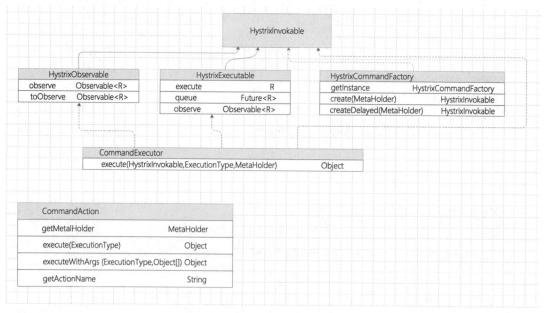

图7-33　本节介绍的类的相关类图

通过代码和类图会发现，上述类结构中使用了设计模式中的命令模式进行设计。其中，HystrixInvokable 是 HystrixCommand 的标记接口，继承了该接口的类都是可以被执行的 HystrixCommand。提供具体方法的接口为 HystrixExecutable，用于同步执行和异步执行命令，HystrixObservable 用于异步回调执行命令，它们分别对应命令模式中的 Command 和 ConcreteCommand。CommandExecutor 会调用 HystrixInvokable 执行命令，相当于命令模式中的 Invoker。HystrixCommandFactory 会生成命令，而 HystrixCommandAspect 相当于命令模式中的客户端情景类 Client。CommandAction 中有 Fallback 方法或者被 @HystrixCommand 注解的远程调用方法，相当于命令模式中的 Receiver。图 7-34 所示为通用命令模式类图。

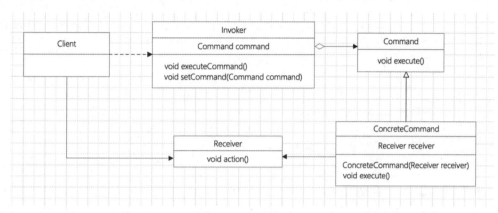

图7-34　通用命令模式类图

7.5.2　断路器逻辑

HystrixCircuitBreaker 是 Hystrix 提供断路器逻辑的核心接口，它通过 HystrixCommandKey（由 @HystrixCommand 的 CommandKey 构造而成）与每一个 HystrixCommand 绑定。在 HystrixCircuitBreaker.Factory 中使用 ConcurrentHashMap 维护了基于 HystrixCommandKey 的 HystrixCircuitBreaker 的单例映射表，以保证具备相同 CommandKey 的 HystrixCommand 对应同一个断路器。

HystrixCircuitBreaker 提供的接口如下所示。

```
// HystrixCircuitBreaker.java

public interface HystrixCircuitBreaker {
// 是否允许命令执行
    boolean allowRequest();
// 断路器是否打开
    boolean isOpen();
// 在半开状态时作为命令执行成功反馈
    void markSuccess();
```

HystrixCircuitBreaker 有两个默认实现：一个是 NoOpCircuitBreaker，顾名思义，即空实现，不会发挥任何断路器的功能；另一个为 HystrixCircuitBreakerImpl，为断路器的真正实现。

HystrixCircuitBreakerImpl 断路器具体实现

在 HystrixCircuitBreakerImpl 中定义了 3 种状态——关闭、打开和半开，与前文介绍的断路器的 3 种状态相对应，如下所示。

```
// HystrixCircuitBreakerImpl.java

public static class HystrixCircuitBreakerImpl implements HystrixCircuitBreaker {
    private final HystrixCommandProperties properties;
```

```java
    private final HystrixCommandMetrics metrics;
    private AtomicBoolean circuitOpen = new AtomicBoolean(false);
    private AtomicLong circuitOpenedOrLastTestedTime = new AtomicLong();

    protected HystrixCircuitBreakerImpl(HystrixCommandKey key, HystrixCommandGroupKey commandGroup, HystrixCommandProperties properties, HystrixCommandMetrics metrics) {
        this.properties = properties;
        this.metrics = metrics;
    }

    public void markSuccess() {
        if (this.circuitOpen.get() && this.circuitOpen.compareAndSet(true, false)) {
            this.metrics.resetStream();
        }

    }

    public boolean allowRequest() {
        if ((Boolean)this.properties.circuitBreakerForceOpen().get()) {// 打开
            return false;
        } else if ((Boolean)this.properties.circuitBreakerForceClosed().get()) {// 关闭
            this.isOpen();
            return true;
        } else {// 半开
            return !this.isOpen() || this.allowSingleTest();
        }
    }

    public boolean allowSingleTest() {
        long timeCircuitOpenedOrWasLastTested = this.circuitOpenedOrLastTestedTime.get();
        return this.circuitOpen.get() && System.currentTimeMillis() > timeCircuitOpenedOrWasLastTested + (long)(Integer)this.properties.circuitBreakerSleepWindowInMilliseconds().get() && this.circuitOpenedOrLastTestedTime.compareAndSet(timeCircuitOpenedOrWasLastTested, System.currentTimeMillis());
    }

    public boolean isOpen() {
        if (this.circuitOpen.get()) {
            return true;
        } else {
            HystrixCommandMetrics.HealthCounts health = this.metrics.getHealthCounts();
            if (health.getTotalRequests() < (long)(Integer)this.properties.circuitBreakerRequestVolumeThreshold().get()) {
                return false;
            } else if (health.getErrorPercentage() < (Integer)this.properties.circuitBreakerErrorThresholdPercentage().get()) {
                return false;
            } else if (this.circuitOpen.compareAndSet(false, true)) {
                this.circuitOpenedOrLastTestedTime.set(System.currentTimeMillis());
                return true;
            } else {
                return true;
            }
        }
    }
```

```
    }
}
```

断路器强制打开和关闭的相关配置可以通过配置中心动态修改，这样就可以人为干预断路器的状态，方便调试。断路器打开的时候会记录打开时间，用于判断断路器是否打开，通过该时间与配置中的 circuitBreakerSleepWindowInMilliseconds 重置时间结合判断：在断路器打开一段时间后（重置时间结束），允许尝试执行命令，以检查远程调用是否恢复为可使用的状态。在第一次发现重置时间结束时，会尝试将断路器的状态从打开修改为半开，方便在命令执行成功或者失败后关闭断路器或重新打开断路器。

markSuccess 方法在命令执行成功后进行调用，将断路器从半开状态转换为关闭状态，同时重置断路器在 HystrixCommandMetrics 中的统计记录和设置断路器打开时间为 −1，即关闭断路器。

■ HystrixCommandMetrics 统计命令执行情况

断路器通过向 HystrixCommandMetrics 中的请求执行统计 Observable 发起订阅来完成断路器自动打开的相关逻辑。

HystrixCommandMetrics 统计了同一 HystrixCommand 请求的指标数据，包括链路健康统计流 HealthCountsStream。HealthCountsStream 使用滑动窗口的方式对各项数据的 HealthCounts 进行统计，在一个滑动窗口时间中又划分了若干个 Bucket（滑动窗口时间与 Bucket 成整数倍关系），滑动窗口的移动是以 Bucket 为单位的，每个 Bucket 仅统计该时间间隔内的请求数据。最后按照滑动窗口的大小对每个 Bucket 中的统计数据进行聚合，得到周期时间内的统计数据 HealthCounts。以下是 HealthCounts 中统计的数据项。

```java
// HealthCounts.java

public static class HealthCounts {
// 执行总次数
    private final long totalCount;
// 执行失败次数
    private final long errorCount;
// 执行失败百分比
    private final int errorPercentage;
    ...
}
```

Hystrix 使用 rx 中的 Observable#window 实现滑动窗口，通过 rx 中单线程的无锁特性保证计数变更时的线程安全，后台线程创建新 Bucket，以避免并发情况。

下面是 HealthCountsStream 的父类创建滑动窗口的相关代码。

```java
// BucketedRollingCounterStream.java

public abstract class BucketedRollingCounterStream<Event extends HystrixEvent, Bucket, Output>
        extends BucketedCounterStream<Event, Bucket, Output> {
    private Observable<Output> sourceStream;
    private final AtomicBoolean isSourceCurrentlySubscribed = new AtomicBoolean(false);

    protected BucketedRollingCounterStream(HystrixEventStream<Event> stream, final int numBuckets, int bucketSizeInMs, Func2<Bucket, Event, Bucket> appendRawEventToBucket, final
```

```java
    Func2<Output, Bucket, Output> reduceBucket) {
        super(stream, numBuckets, bucketSizeInMs, appendRawEventToBucket);
            Func1<Observable<Bucket>, Observable<Output>> reduceWindowToSummary = new
    Func1<Observable<Bucket>, Observable<Output>>() {
                public Observable<Output> call(Observable<Bucket> window) {
// 合并 20 个 Bucket 数据项
                    return window.scan(BucketedRollingCounterStream.this.getEmptyOutputValue(),
 reduceBucket).skip(numBuckets);
                }
            };
        this.sourceStream = this.bucketedStream.window(numBuckets, 1).flatMap(reduceWindowToSu
mmary).doOnSubscribe(new Action0() {
            public void call() {
                BucketedRollingCounterStream.this.isSourceCurrentlySubscribed.set(true);
            }
        }).doOnUnsubscribe(new Action0() {
            public void call() {
                BucketedRollingCounterStream.this.isSourceCurrentlySubscribed.set(false);
            }
        }).share().onBackpressureDrop();
    }

    public Observable<Output> observe() {
        return this.sourceStream;
    }

    boolean isSourceCurrentlySubscribed() {
        return this.isSourceCurrentlySubscribed.get();
    }
}
```

在上述代码中，window 定义了每发送一次数据（此时一个数据项将会被从滑动窗口中移除，以及创建一个新的 Bucket 用于统计）都会聚合 numBuckets 个数据项，即整个滑动窗口的统计数据集合，numBuckets 一般为 20 个。

合并数据项的代码在 HealthCounts#plus 中。该方法用于将 Bucket 中的统计数据合并为 HealthCounts 发出。具体代码如下所示。

```java
// HealthCounts.java

public HystrixCommandMetrics.HealthCounts plus(long[] eventTypeCounts) {
    long updatedTotalCount = this.totalCount;
    long updatedErrorCount = this.errorCount;
// 成功次数
    long successCount = eventTypeCounts[HystrixEventType.SUCCESS.ordinal()];
// 失败次数
    long failureCount = eventTypeCounts[HystrixEventType.FAILURE.ordinal()];
// 超时次数
    long timeoutCount = eventTypeCounts[HystrixEventType.TIMEOUT.ordinal()];
// 请求线程失败次数
    long threadPoolRejectedCount = eventTypeCounts[HystrixEventType.THREAD_POOL_
REJECTED.ordinal()];
// 请求信号量失败次数
    long semaphoreRejectedCount = eventTypeCounts[HystrixEventType.SEMAPHORE_REJECTED.
```

```
  ordinal()];
// 执行总次数
    updatedTotalCount += successCount + failureCount + timeoutCount + threadPoolRejectedCount
+ semaphoreRejectedCount;
// 打开错误次数
    updatedErrorCount += failureCount + timeoutCount + threadPoolRejectedCount +
semaphoreRejectedCount;
    return new HystrixCommandMetrics.HealthCounts(updatedTotalCount, updatedErrorCount);
}
```

上述代码将每个 Bucket 数据统计并合并成一个完整的 HealthCounts 发送，统计的是近 10s 内的命令执行情况。

在 BucketedRollingCounterStream 的父类 BucketedCounterStream 中，有如下规定。

healthCountBucketSizeInMs 间隔后发送一次数据，同时初始化 numBuckets 个 Bucket 用于统计，代码如下所示。

```
// BucketedCounterStream.java

final List<Bucket> emptyEventCountsToStart = new ArrayList();

for(int i = 0; i < numBuckets; ++i) {
    emptyEventCountsToStart.add(this.getEmptyBucketSummary());
}

this.bucketedStream = Observable.defer(new Func0<Observable<Bucket>>() {
    public Observable<Bucket> call() {
        return inputEventStream.observe().window((long)bucketSizeInMs, TimeUnit.MILLISECONDS).flatMap(BucketedCounterStream.this.reduceBucketToSummary).startWith(emptyEventCountsToStart);
    }
});
```

HystrixCircuitBreaker 通过对 HealthCountsStream 进行订阅，监控链路健康情况，在一定条件下打开断路器。在固定间隔时间（默认为 500ms），滑动窗口中聚合而成的表示链路健康统计数据的 HealthCounts 会被用来检查是否打开断路器。只有在周期内（10s）请求的总数超过一定的阈值且执行失败的百分比超过 circuitBreakerErrorThresholdPercentage 时，断路器才会被打开。

触发 HystrixCommandMetrics 的统计命令执行结果主要发生在 AbstractCommand 中，如下所示。

```
// AbstractCommand.java

private void handleCommandEnd(boolean commandExecutionStarted) {
    ...
// 统计命令执行结果
    if (this.executionResultAtTimeOfCancellation == null) {
        this.metrics.markCommandDone(this.executionResult, this.commandKey, this.threadPoolKey,
commandExecutionStarted);
    } else {
        this.metrics.markCommandDone(this.executionResultAtTimeOfCancellation, this.commandKey,
this.threadPoolKey, commandExecutionStarted);
    }
}
```

```
    ...
}
```

AbstractCommand 在命令执行结束后的回调方法中,通过 HystrixCommandMetrics 统计相关命令的执行结果,这其中主要通过 HystrixCommandCompletion 数据类对命令执行结束后的事件流进行统计,事件类型由 HystrixEventType 定义。

HystrixCommandCompletion 由 HystrixCommandCompletionStream 进行管理,最终在 HealthCountsStream 中用于统计一段时间内的链路健康情况。

第 8 章

Spring Cloud Gateway 服务网关

在学习完前面的知识后,微服务架构在读者脑海中已经初具雏形。但还有一些问题,比如不同的微服务通常有不同的网络地址,客户端在访问这些微服务时必须记住几十甚至几百个地址,这对客户端来说太复杂也难以维护。这个问题可以使用 Gateway 服务网关来解决。

本章的主要内容如下。

1. 认识 Spring Cloud Gateway。
2. 实现服务网关。
3. 路由规则。
4. 过滤器。
5. 网关限流。
6. 源码解析。

8.1 认识 Spring Cloud Gateway

Spring Cloud Gateway 是 Spring Cloud 的一个全新项目，该项目是基于 Spring Framework 5、Spring Boot 2.0 和 Project Reactor 等技术开发的网关，旨在为微服务架构提供一种简单有效的、统一的 API 路由管理方式。

Spring Cloud Gateway 作为 Spring Cloud 生态系统中的网关，其目标是替代 Zuul。在 Spring Cloud 2.0 以上的版本中，没有对 Zuul 2.0 以上最新高性能版本进行集成，仍然使用的是 Zuul 2.0 之前的非 Reactor 模式的老版本。为了提升网关的性能，Spring Cloud Gateway 是基于 WebFlux 框架实现的，而 WebFlux 框架底层使用了高性能的 Reactor 模式通信框架 Netty。

Spring Cloud Gateway 旨在提供一种简单而有效的途径来转发请求，并为它们提供横切关注点，例如安全性、监控指标和弹性。

Spring Cloud Gateway 具有以下特征。

（1）基于 Java 8 编码。

（2）支持 Spring Framework 5。

（3）支持 Spring Boot 2.0。

（4）支持动态路由。

（5）支持内置到 SpringHandler 映射中的路由匹配。

（6）支持基于 HTTP 请求的路由匹配（Path、Method、Header、Host 等）。

（7）过滤器作用于匹配的路由。

（8）过滤器可以修改下游 HTTP 请求和 HTTP 响应（增加／修改头部、增加／修改请求参数、改写请求路径等）。

（9）通过 API 或配置驱动。

（10）支持 Spring Cloud Discovery Client 配置路由，与服务注册和发现配合使用。

在 Finchley 正式版发布之前，Spring Cloud 推荐的网关是 Netflix 提供的 Zuul（这里指的都是 Zuul 1.x，一个基于阻塞 I/O 的 API Gateway）。与 Zuul 相比，Spring Cloud Gateway 建立在 Spring Framework 5、Project Reactor 和 Spring Boot 2.0 的基础之上，使用非阻塞 API。Spring Cloud Gateway 还支持 WebSocket，并且与 Spring 紧密集成，可以使用户拥有更好的开发体验。Zuul 基于 Servlet 2.5，使用阻塞架构，不支持任何长连接，如 WebSocket。Zuul 的设计模式和 Nginx 较像，每次 I/O 操作都是从工作线程中选择一个执行，请求线程被阻塞直到工作线程完成，但是差别是 Nginx 用 C++ 实现，Zuul 用 Java 实现，而 JVM 本身会有第一次加载较慢的情况，使得 Zuul 的性能相对较差。Zuul 已经发布了 Zuul 2.x，基于 Netty、非阻塞、支持长连接，但 Spring Cloud 目前还没有集成。Zuul 2.x 的性能肯定会较 Zuul 1.x 有较大提升。在性能方面，根据官方提供的基准测试，Spring Cloud Gateway 的 RPS（每秒请求数）是 Zuul 的 1.6 倍。综合来说，Spring Cloud Gateway 在提供的功能和实际性能方面，表现都很优异。

8.1.1 微服务网关概述

在学习完前面的知识后,微服务架构在读者脑海中已经初具雏形。但还有一些问题,比如不同的微服务通常有不同的网络地址,客户端在访问这些微服务时必须记住几十甚至几百个地址,这对客户端来说太复杂也难以维护。未使用微服务网关的情况如图 8-1 所示。

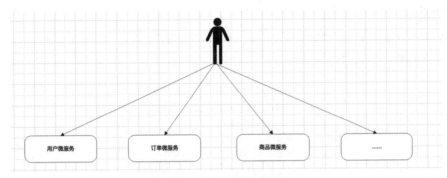

图8-1 未使用微服务网关

如果让客户端直接与各个微服务通信,可能会有很多问题,如下所示。

(1)客户端会请求多个不同的服务,需要维护不同的请求地址,增加了开发的难度。

(2)在某些场景下存在跨域请求的问题。

(3)加大了身份认证的难度,每个微服务都需要独立认证。

因此,需要一个微服务网关,介于客户端与服务器之间,这样所有的外部请求都会先经过微服务网关。客户端只需要与网关交互,只知道一个网关地址即可,这样简化了开发。使用微服务网关的情况如图 8-2 所示。

使用微服务网关有以下优点。

(1)易于监控。

(2)易于认证。

(3)减少了客户端与各个微服务之间的交互次数。

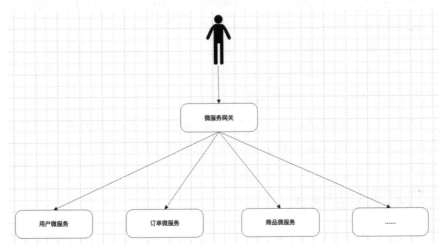

图8-2 使用微服务网关

8.1.2 微服务网关工作流程

微服务网关通常是一个服务器，是系统对外的唯一入口。微服务网关封装了系统内部架构。为每个客户端（Client）提供一个定制的 API。使用微服务网关的核心是所有的客户端和服务端（Server）都通过统一的网关接入微服务，在网关层处理所有的非业务功能。通常，网关也提供 REST/HTTP 的访问 API，服务端可注册和管理服务。

网关具有身份验证、监控、负载均衡、缓存、请求分片与管理、静态响应处理等功能。当然，网关最主要的功能还是与外界联系。微服务网关工作流程如图 8-3 所示。

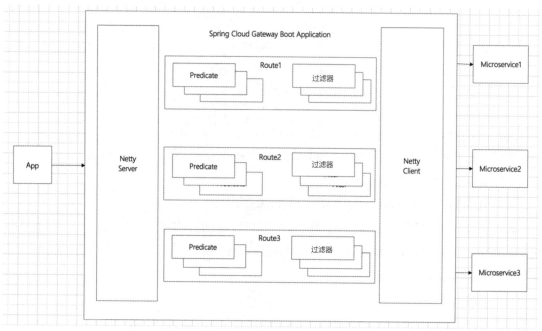

图8-3 微服务网关工作流程

路由（Route）是网关最基础的部分之一。路由信息由一个 ID、一个目的 URL、一组断言工厂和一组过滤器组成。如果断言为真，则说明请求的 URL 和配置的路由匹配。

断言（Predicate）是 Java 8 中的断言函数。Spring Cloud Gateway 中的断言函数输入类型是 Spring Framework 5 中的 ServerWebExchange。Spring Cloud Gateway 中的断言函数允许开发者定义匹配来自 HttpRequest 的任何信息，比如请求头和请求参数等。

过滤器（Filter）是一个标准的 Spring Web Filter。Spring Cloud Gateway 中的过滤器分为两种类型，分别是 GatewayFilter 和 GlobalFilter。过滤器可以对请求和响应进行处理。

实现服务网关

以 Eureka 为注册中心，实现商品微服务，端口号分别设置为 9000 和 9002。实现微服务网关，通过微服务网关调用商品微服务，端口号设置为 8080。

8.2.1 创建子工程——服务网关

使用 IntelliJ IDEA 创建子工程——服务网关（gateway-server），如图 8-4 所示。

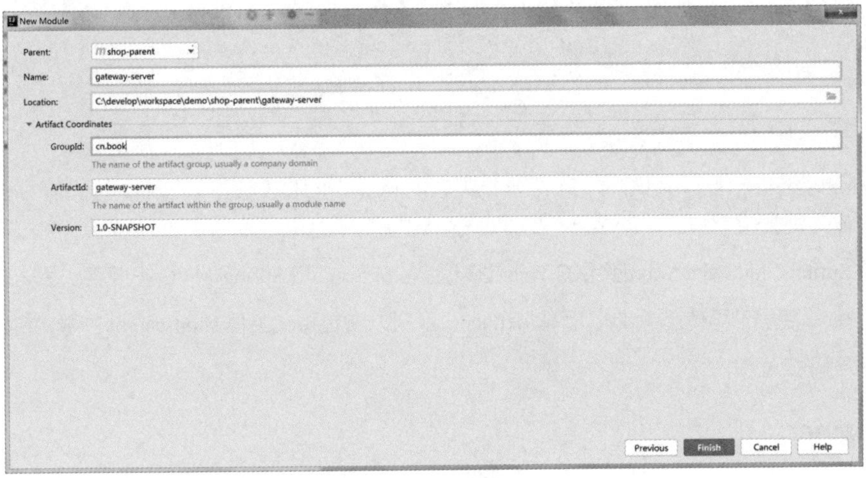

图8-4 服务网关

项目结构如图 8-5 所示。

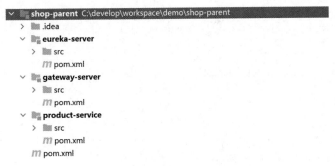

图8-5 项目结构

8.2.2 坐标依赖

在服务网关 gateway-server 中添加 Gateway 依赖，如下所示。

```
<dependency>
    <groupId>org.springframework.cloud</groupId>
    <artifactId>spring-cloud-starter-gateway</artifactId>
</dependency>
```

8.2.3 工程改造

如果此时启动此服务网关会报错，如下所示。

```
***************************
APPLICATION FAILED TO START
***************************

Description:

Parameter 0 of method modifyRequestBodyGatewayFilterFactory in org.springframework.cloud.
gateway.config.GatewayAutoConfiguration required a bean of type 'org.springframework.http.
codec.ServerCodecConfigurer' that could not be found.

Action:

Consider defining a bean of type 'org.springframework.http.codec.ServerCodecConfigurer' in
 your configuration.
```

原因是 Spring Cloud Gateway 使用的 Web 框架为 WebFlux，和 Spring MVC 不兼容。引入的限流组件是 Hystrix。Redis 底层不再使用 Jedis，而是 Lettuce。解决办法是将父工程 shop-parent 中导入的 spring-boot-starter-web 移到子工程中。

■ shop-parent

父工程 shop-parent 的 pom.xml 文件中的坐标依赖部分内容如下所示。

```xml
<dependencies>
  <dependency>
    <groupId>org.springframework.boot</groupId>
    <artifactId>spring-boot-starter-logging</artifactId>
  </dependency>
  <dependency>
    <groupId>org.springframework.boot</groupId>
    <artifactId>spring-boot-starter-test</artifactId>
    <scope>test</scope>
  </dependency>
  <dependency>
    <groupId>org.projectlombok</groupId>
    <artifactId>lombok</artifactId>
    <version>1.18.4</version>
    <scope>provided</scope>
  </dependency>
</dependencies>
```

■ product-service

子工程 product-service 的 pom.xml 文件中的坐标依赖部分内容如下所示。

```xml
<dependencies>
  <dependency>
    <groupId>org.springframework.boot</groupId>
    <artifactId>spring-boot-starter-web</artifactId>
  </dependency>
  <dependency>
    <groupId>mysql</groupId>
```

```xml
      <artifactId>mysql-connector-java</artifactId>
      <version>5.1.32</version>
    </dependency>
    <dependency>
      <groupId>org.springframework.boot</groupId>
      <artifactId>spring-boot-starter-data-jpa</artifactId>
    </dependency>
    <dependency>
      <groupId>org.springframework.cloud</groupId>
      <artifactId>spring-cloud-starter-netflix-eureka-client</artifactId>
    </dependency>
</dependencies>
```

■ eureka-server

子工程 eureka-server 的 pom.xml 文件中的坐标依赖部分内容如下所示。

```xml
<dependencies>
  <dependency>
    <groupId>org.springframework.boot</groupId>
    <artifactId>spring-boot-starter-web</artifactId>
  </dependency>
  <dependency>
    <groupId>org.springframework.cloud</groupId>
    <artifactId>spring-cloud-starter-netflix-eureka-server</artifactId>
  </dependency>
</dependencies>
```

■ gateway-server

子工程 gateway-server 的 pom.xml 文件中的坐标依赖部分内容如下所示。

```xml
<dependencies>
  <dependency>
    <groupId>org.springframework.cloud</groupId>
    <artifactId>spring-cloud-starter-gateway</artifactId>
  </dependency>
</dependencies>
```

创建子工程 gateway-server 的启动类，如下所示。

```java
@SpringBootApplication
public class GatewayApplication {
  public static void main(String[] args) {
    SpringApplication.run(GatewayApplication.class,args);
  }
}
```

创建子工程 gateway-server 的配置文件 application.yml，如下所示。

```yaml
server:
  port: 8080
spring:
  application:
    name: gateway-server
```

```yaml
    cloud:
      gateway:
        routes:
          - id: product-service
            uri: http://10.211.55.12:9002/
            predicates:
              - Path=/product/**
```

gateway 部分参数介绍如下。

（1）id：自定义的路由 ID，保持唯一。

（2）uri：目标服务地址。

（3）predicates：路由条件。接收一个输入参数，返回一个布尔值。有多种默认方法来将 Predicate 组合成其他复杂的逻辑（如与、或、非）。

（4）filters：过滤规则，后续小节会进行讲解。

如果配置了一个 id 为 product-service 的路由规则，当访问的网关请求地址以 product 开头时，会自动转发到地址 http://10.211.55.12:9002/ 上。

8.2.4 代码测试

重启各个服务后，访问 product-service，结果如图 8-6 所示。

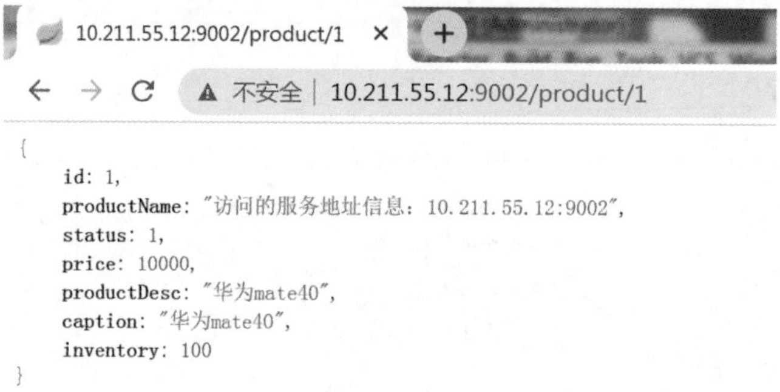

图8-6 访问product-service

访问 gateway-server，结果如图 8-7 所示。

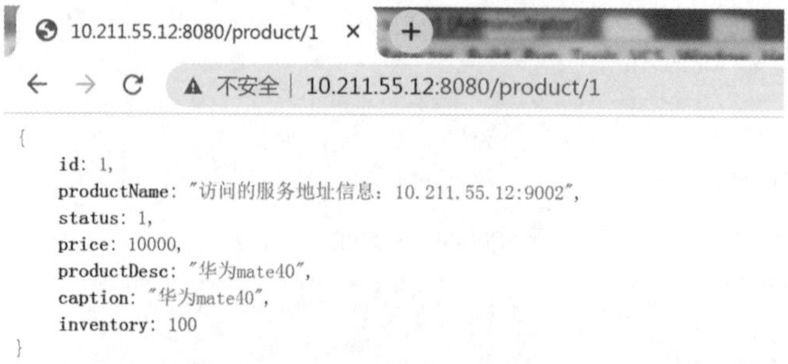

图8-7 访问gateway-server

8.3 路由规则

Spring Cloud Gateway 的功能很强大,前面只是使用断言进行了简单的条件匹配,其实 Spring Cloud Gateway 内置了很多断言功能。在 Spring Cloud Gateway 中,Spring 利用断言的特性实现了各种路由规则,通过 Header、请求参数等作为条件匹配到对应的路由,如图 8-8 所示。

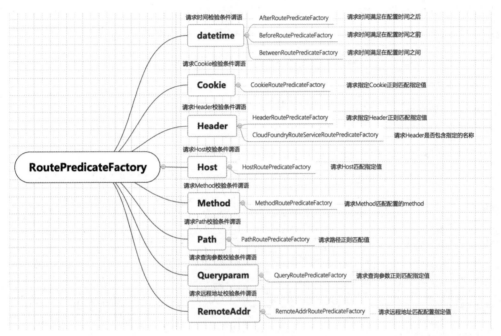

图8-8 断言功能

8.3.1 路由规则概述

在图 8-8 中对各个路由工厂的规则进行了列举,简单地理解:断言用于实现一组路由规则,方便让请求找到对应的路由进行处理。下面对各个路由规则进行介绍。

时间匹配

Spring Cloud Gateway 支持设置一个时间,在请求转发的时候,可以判断是在这个时间之前还是之后进行转发。比如设置只有在 2020 年 1 月 9 日之后才会进行路由转发,在这之前不进行转发,如下所示。

```
server:
  port: 8080
spring:
  application:
    name: gateway-server
  cloud:
    gateway:
```

```yaml
routes:
 - id: ecs-route
   uri: lb://test-ecs
   predicates:
    - Path=/ecs/**
    - After=2020-01-09T00:00:00+08:00[Asia/Shanghai]
 - id: oss-route
   uri: lb://test-oss
   predicates:
    - Path=/oss/**
```

上面配置了 After 属性，表示在这个时间之后才可以进行时间匹配；我们设置的是 2020-01-09，这里的时间是通过 ZonedDateTime 来进行对比的。ZonedDateTime 是 Java 8 中的日期时间，是带时区的日期与时间信息的类。ZonedDateTime 支持通过时区来设置时间，这里使用的时区是：Asia/Shanghai。

上面的代码是指请求路径为 /ecs/，且请求时间在 2020 年 1 月 9 日 0 点 0 分 0 秒之后的所有请求都转发到地址 http://test-ecs/ecs/ 上。+08:00 是指当地时间和 UTC 时间相差 8h，时区为 Asia/Shanghai。

添加完路由规则之后，如果在当地时间 1 月 8 日访问地址 http://localhost:8080/ecs 不会进行转发，如果是 1 月 9 日访问则会自动转发到 http://test-ecs/ecs/。

把上面路由规则中的 After 改为 Before，则会对在某个时间之前的请求都进行转发，如下所示。

```yaml
server:
 port: 8080
spring:
 application:
  name: gateway-server
 cloud:
  gateway:
   routes:
    - id: ecs-route
      uri: lb://test-ecs
      predicates:
       - Path=/ecs/**
       - Before=2020-01-09T00:00:00+08:00[Asia/Shanghai]
    - id: oss-route
      uri: lb://test-oss
      predicates:
       - Path=/oss/**
```

有了时间前与时间后，那么时间之间的 Between 实现代码如下所示。

```yaml
server:
 port: 8080
spring:
 application:
  name: gateway-server
 cloud:
  gateway:
   routes:
    - id: ecs-route
      uri: lb://test-ecs
      predicates:
```

```
          - Path=/ecs/**
          - Between=2020-01-01T00:00:00+08:00[Asia/Shanghai],2020-01-09T00:00:00+08:00[Asia/Shanghai]
        - id: oss-route
          uri: lb://test-oss
          predicates:
          - Path=/oss/**
```

上面的 Between 配置表示在两个时间之间即可进行路由转发，否则不可进行路由转发。

■ Cookie 匹配

Gateway 的 Cookie 匹配接收两个参数：一个是 Cookie name，另一个是正则表达式。其路由规则就是通过获取对应的 Cookie name 值和正则表达式去匹配路由，如果匹配上就会执行路由，否则不执行，如下所示。

```
server:
  port: 8080
spring:
  application:
    name: gateway-server
  cloud:
    gateway:
      routes:
        - id: ecs-route
          uri: lb://test-ecs
          predicates:
          - Path=/ecs/**
          - Cookie=userCode,sxd782wc
        - id: oss-route
          uri: lb://test-oss
          predicates:
          - Path=/oss/**
```

不带 Cookies 访问，返回 404 异常。带上 Cookies 访问，且 userCode 为 sxd782wc，则正常返回。

■ Header 匹配

Header 匹配和 Cookie 匹配一样，也接收两个参数，一个是 Header 中的属性名称，另一个是正则表达式。如果属性值和正则表达式匹配则执行。路由规则如下所示。

```
server:
  port: 8080
spring:
  application:
    name: gateway-server
  cloud:
    gateway:
      routes:
        - id: ecs-route
          uri: lb://test-ecs
          predicates:
          - Path=/ecs/**
          - Header=token,^(?!\d+$)[\da-zA-Z]+$
        - id: oss-route
```

```yaml
      uri: lb://test-oss
      predicates:
      - Path=/oss/**
```

上面的路由规则就是请求头要有 token 属性，并且其值必须为由数字和字母组成的正则表达式。

■ **Host 匹配**

Host 匹配接收一组参数和一组匹配的域名列表下面这个 Host 匹配模板是一个 Ant 分隔的模板，用 "."符号作为分隔符，它通过参数中的主机地址作为路由规则。

```yaml
server:
 port: 8080
spring:
 application:
  name: gateway-server
 cloud:
  gateway:
   routes:
   - id: ecs-route
     uri: lb://test-ecs
     predicates:
     - Host=**.xxxx.com
   - id: oss-route
     uri: lb://test-oss
     predicates:
     - Path=/oss/**
```

如果请求 Host 符合要求，则此路由匹配，例如 www.xxxx.com 或 map.xxxx.com。

■ **请求方式匹配**

通过请求方式 POST、GET、PUT、DELETE 等进行路由匹配，如下所示。

```yaml
server:
 port: 8080
spring:
 application:
  name: gateway-server
 cloud:
  gateway:
   routes:
   - id: ecs-route
     uri: lb://test-ecs
     predicates:
     - Method=GET // 或者 POST、PUT
   - id: oss-route
     uri: lb://test-oss
     predicates:
     - Path=/oss/**
```

■ **请求路径匹配**

请求路径（Path）匹配接收一个匹配路径的参数来判断是否进行路由，如下所示。

```
server:
  port: 8080
spring:
  application:
    name: gateway-server
  cloud:
    gateway:
      routes:
      - id: ecs-route
        uri: lb://test-ecs
        predicates:
        - Path=/ecs/{segment}
      - id: oss-route
        uri: lb://test-oss
        predicates:
        - Path=/oss/**
```

如果请求路径符合要求，则此路由匹配，例如 /ecs/1 或 /ecs/bar。

■ **请求参数匹配**

请求参数（Query）匹配支持传入两个参数，一个是属性名，另一个是属性值，其中属性值可以是正则表达式，如下所示。

```
server:
  port: 8080
spring:
  application:
    name: gateway-server
  cloud:
    gateway:
      routes:
      - id: ecs-route
        uri: lb://test-ecs
        predicates:
        - Query=token
      - id: oss-route
        uri: lb://test-oss
        predicates:
        - Query=token, \d+
```

第 1 个路由规则：只有属性名 token，表示只要有 token 参数即可路由，不管其值是什么。

第 2 个路由规则：不但要有属性名 token，属性值还得是整数才能路由。

■ **请求 IP 地址匹配**

通过设置 IP 地址属于某个区间号段的请求才会路由，RemoteAddr 接受 CIDR 符号（IPv4 或 IPv6）的列表（最小大小为 1），例如 192.168.0.1/16，其中 192.168.0.1 是 IP 地址，16 是子网掩码。请求 IP 地址匹配代码如下所示。

```
predicates:
  - RemoteAddr=192.168.1.1/24
```

如果请求的远程地址是 192.168.1.11，则此路由将匹配。

■ 组合使用

各种断言同时存在于同一个路由规则中时，请求必须同时满足所有的条件才能被这个路由匹配。

```
predicates:
  - Host=**.xxxx.com
  - Path=/ecs/test
  - Method=GET
  - Query=token
  - After=2020-01-9T06:06:06+08:00[Asia/Shanghai]
```

8.3.2 动态路由

在 8.2 节中已经实现了微服务网关的路由转发，但是把转发的 uri 地址硬编码到代码中，会存在很多问题，比如应用场景有局限和无法动态调整，那怎么解决呢？这就需要动态路由，即自动从注册中心获取服务列表并访问。

■ 添加注册中心依赖

子工程 eureka-server 的 pom.xml 文件中的坐标依赖部分内容如下所示。

```xml
<dependency>
    <groupId>org.springframework.cloud</groupId>
    <artifactId>spring-cloud-starter-netflix-eureka-client</artifactId>
</dependency>
```

■ 配置动态路由

修改 application.yml 配置文件，添加 Eureka 注册中心的相关配置，并修改访问映射的 uri 为服务名称，如下所示。

```yaml
server:
  port: 8080 # 服务端口
spring:
  application:
    name: gateway-server # 指定服务名称
  cloud:
    gateway:
      routes:
        - id: product-service
          uri: lb://product-service
          predicates:
            - Path=/product/**
eureka: # 配置 Eureka
  client:
    service-url:
```

```
        defaultZone: http://10.211.55.12:9000/eureka/ # 多个 Eureka 服务端之间用逗号隔开
    instance:
      prefer-ip-address: true # 使用 IP 地址注册
      instance-id: ${spring.cloud.client.ip-address}:${server.port} # 向注册中心注册服务 id
```

uri 以"lb://"开头（lb 表示从注册中心获取服务），后面接的就是需要转发到的服务名称，这里转发到商品微服务。

■ 代码测试

重启各个服务后，访问 gateway-server，结果如图 8-9 所示。

图8-9 访问gateway-server

8.3.3 重写转发路径

在 Spring Cloud Gateway 中，路由转发是直接将匹配的路由 Path 拼接到映射路径（uri）之后，而在微服务开发中往往没有那么便利。这里可以通过 RewritePath 机制来进行路径重写。

■ 工程改造

修改 application.yml，将匹配路径改为 /product-service/**，如下所示。

```
server:
  port: 8080 # 服务端口
spring:
  application:
    name: gateway-server # 指定服务名称
  cloud:
    gateway:
      routes:
        - id: product-service
          uri: lb://product-service
          predicates:
            - Path=/product-service/**
eureka: # 配置 Eureka
  client:
    service-url:
      defaultZone: http://10.211.55.12:9000/eureka/ # 多个 Eureka 服务端之间用逗号隔开
  instance:
    prefer-ip-address: true # 使用 IP 地址注册
```

    instance-id: ${spring.cloud.client.ip-address}:${server.port} # 向注册中心注册服务 id

重新启动网关，在浏览器中访问 http://10.211.55.12:8080/product-service/product/1，会抛出 404 异常，如图 8-10 所示。

图8-10　访问结果

这是由于路由规则默认转发到商品微服务 http://10.211.55.12:9002/product-service/product/1 路径上，而商品微服务又没有 product-service 对应的映射配置。

■ 添加 RewritePath 重写转发路径

修改 application.yml，添加重写规则，如下所示。

```yaml
server:
  port: 8080 # 服务端口
spring:
  application:
    name: gateway-server # 指定服务名称
  cloud:
    gateway:
      routes:
        - id: product-service
          uri: lb://product-service
          predicates:
            - Path=/product-service/**
          filters:
            - RewritePath=/product-service/(?<segment>.*),/$\{segment}
eureka: # 配置 Eureka
  client:
    service-url:
      defaultZone: http://10.211.55.12:9000/eureka/ # 多个 Eureka 服务端之间用逗号隔开
  instance:
    prefer-ip-address: true # 使用 IP 地址注册
    instance-id: ${spring.cloud.client.ip-address}:${server.port} # 向注册中心注册服务 id
```

重启各个服务后，访问 gateway-server，结果如图 8-11 所示。

图8-11　使用RewritePath后访问gateway-server

上述代码中通过 RewritePath 重写转发的 uri，将 /product-service/(?<segment>.*) 重写为 {segment}，然后转发到商品微服务。比如在浏览器上访问 http://10.211.55.12:8080/product-service/product/1，此时会将请求转发到 http://10.211.55.12:9002/product/1。值得注意的是在 YML 文件中 $ 要写成 $\。

Spring Cloud Gateway 也提供了自动根据微服务名称进行路由转发的配置，配合服务发现一起使用可以简化配置，如下所示。

```yaml
server:
  port: 8080 # 服务端口
spring:
  application:
    name: gateway-server # 指定服务名称
  cloud:
    gateway:
      discovery:
        locator:
          lower-case-service-id: true # 微服务名称以小写形式呈现
          enabled: true # 开启根据微服务名称自动转发
eureka: # 配置 Eureka
  client:
    service-url:
      defaultZone: http://10.211.55.12:9000/eureka/ # 多个 Eureka 服务端之间用逗号隔开
  instance:
    prefer-ip-address: true # 使用 IP 地址注册
    instance-id: ${spring.cloud.client.ip-address}:${server.port} # 向注册中心注册服务 id
```

重启各个服务后，访问 gateway-server，结果如图 8-12 所示。

图8-12　根据微服务名称进行路由转发

8.4　过滤器

Gateway 的一个重要功能就是实现请求的鉴权，而这个功能往往是通过网关提供的过滤器来实现的。在 8.3.3 小节中的重写转发路径也是使用过滤器实现的。接下来一起来研究一下 Gateway 中的过滤器。

8.4.1　过滤器基础

所有开源框架实现过滤器的模式都是大同小异的，即通过一种类似职责链的模式。传统的职责链模式中的事件会一直传递，直到有一个处理对象接手，而过滤器和传统的职责链有点不同，它更像是足球队开场握手一样,所有队员一字排开,你要从头到尾依次和所有球员握手。

Gateway 中的过滤器也采用一样的模式,它们经过优先级的排列,所有网关调用请求从最高优先级的过滤器开始,一路走到头,直到被最后一个过滤器处理。

不同于 Spring Cloud 上一代网关组件 Zuul 里对过滤器的 Pre 和 Post 的定义,Gateway 是通过 Filter 中的代码来实现类似 Pre 和 Post 的功能的。Pre 和 Post 是指当前过滤器的执行阶段,Pre 是在下一个过滤器之前被执行,可利用这种过滤器实现身份验证、在集群中选择请求的微服务、记录调试信息等;Post 是在过滤器执行过后再执行,这种过滤器可用来为响应添加标准的 HTTP Header、收集统计信息和指标、将响应从微服务发送给客户端等。使用者在 GatewayFilter 中也可以同时定义 Pre 和 Post 执行逻辑。多个过滤器的执行流程如图 8-13 所示。

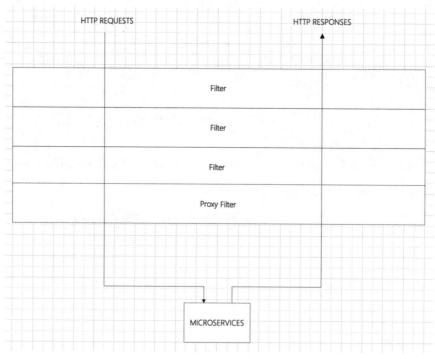

图8-13 多个过滤器的执行流程

Spring Cloud Gateway 的过滤器从其作用范围可分为两种:GatewayFilter、GlobalFilter。GatewayFilter 应用于单个路由或者一个分组的路由上,GlobalFilter 应用于所有的路由上。

8.4.2 局部过滤器

局部过滤器可以对访问的 url 进行过滤和切面处理。在 Spring Cloud Gateway 中以 GatewayFilter 的形式内置了很多不同类型的局部过滤器。这里将 Spring Cloud Gateway 内置的所有局部过滤器整理到了一张表格中,虽然不是很详细,但能作为速览使用,如表 8-1 所示。

表 8-1 局部过滤器

局部过滤器	作用	参数
AddRequestHeader	为原始请求添加 Header	Header 的名称及值
AddRequestParameter	为原始请求添加请求参数	参数名称及值
AddResponseHeader	为原始响应添加 Header	Header 的名称和值
DedupeResponseHeader	剔除响应头中重复的值	需要去重的 Header 名称及去重策略
Hystrix	为路由引入 Hystrix 的断路器保护	HystrixCommand 的名称
FallbackHeaders	为 FallbackUri 的请求头中添加具体的异常信息	Header 的名称
PrefixPath	为原始请求路径添加前缀	前缀名称
PreserveHostHeader	为请求添加一个 PreserveHostHeader=true 的属性，路由过滤器会检查该属性以决定是否要发送原始的 Host	无
RequestRateLimiter	用于对请求限流，限流算法为令牌桶	keyResolve、rateLimiter、statusCode、denyEmptyKey、emptyKeyStatus
RedirectTo	将原始请求重定向到指定 URL	HTTP 状态码及重定向 URL
RemoveHopByHopHeadersFilter	为原始请求删除 IETF 规定的一系列 Header	默认就会启用，可以通过配置指定仅删除哪些 Header
RemoveRequestHeader	为原始请求删除某个 Header	Header 的名称
RemoveResponseHeader	为原始响应删除某个 Header	Header 的名称
RewritePath	重写原始的请求路径	原始路径正则表达式以及重写后路径的正则表达式
RewriteResponseHeader	重写原始响应中的某个 Header	Header 的名称、值的正则表达式、重写后的值
SaveSession	在转发请求之前，强制执行 webSession Save 操作	无
SecureHeader	为原始响应添加一系列起安全作用的响应头	无，支持修改这些安全响应头的值
SetPath	修改原始的请求路径	修改后的路径
SetResponseHeader	修改原始响应中某个 Header 的值	Header 的名称、修改后的值
SetStatus	修改原始响应的状态码	HTTP 状态码，可以是数字，也可以是字符串
StripPrefix	用于截断原始请求的路径	使用数字表示要截断的路径的数量
Retry	针对不同的响应进行重试	retries、statuses、methods、series
RequestSize	设置允许接收最大请求包的大小。如果请求包大小超过设置的值，则返回 413 Payload Too Large	请求包大小，单位为字节，默认值为 5MB
ModifyRequestBody	在转发请求之前修改原始请求体内容	修改后的请求体内容

每个局部过滤器都对应一个实现类，并且这些类的名称必须以 GatewayFilterFactory 结尾，这是 Spring Cloud Gateway 的一个约定，例如 AddRequestHeader 对应的实现类为 AddRequestHeaderGatewayFilterFactory，如图 8-14 所示。对于这些过滤器的使用方式可以参考官方文档。

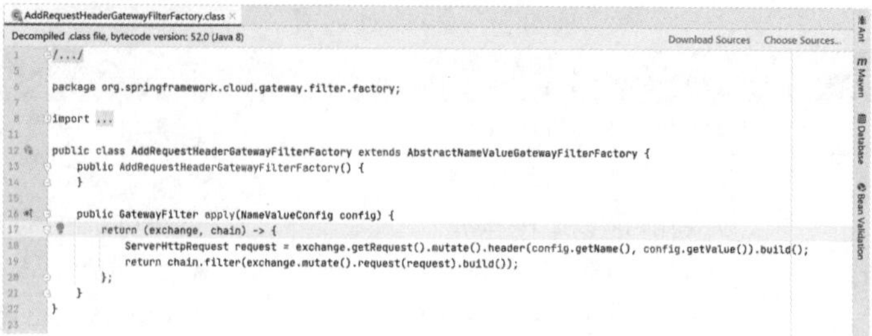

图8-14 AddRequestHeaderGatewayFilterFactory

8.4.3 全局过滤器

全局过滤器（GlobalFilter）作用于所有路由，Spring Cloud Gateway 定义了 GlobalFilter 接口，用户可以自定义实现自己的全局过滤器。通过全局过滤器可以实现对权限的统一校验、安全性验证等功能，并且全局过滤器也是程序员使用比较多的过滤器。

Spring Cloud Gateway 内部也是通过一系列的内置全局过滤器对整个路由转发进行处理的，如图 8-15 所示。

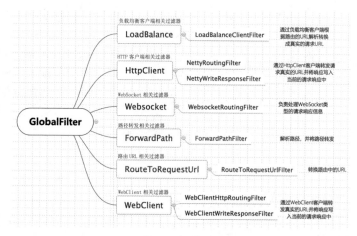

图8-15 全局过滤器

内置的全局过滤器已经可以完成大部分的功能，但是对于企业开发的一些业务功能处理，还是需要使用者自己编写过滤器来实现的。下面通过代码自定义一个过滤器，来完成统一的权限校验。

■ 鉴权逻辑

需要开发的鉴权逻辑如下所示。

（1）当客户端第一次请求服务时，服务端对用户进行信息认证（登录凭证）。

（2）认证通过，将用户信息加密形成 token，返回给客户端，作为登录凭证。

（3）以后每次请求，客户端都携带认证的 token。

（4）服务端对 token 进行解密，判断是否有效。

需要开发的鉴权逻辑流程如图 8-16 所示。

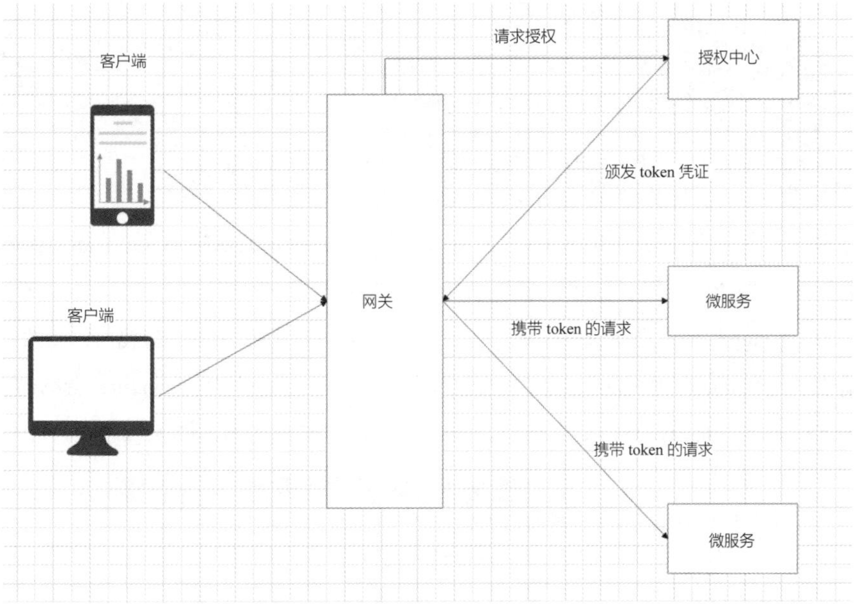

图8-16 鉴权逻辑

如图 8-16 所示，验证用户是否已经登录鉴权的过程可以在网关统一检验。检验的标准就是请求中是否携带 token 凭证以及 token 的正确性。

■ 代码实现

下面自定义一个全局过滤器，校验所有请求的请求参数中是否包含 token，如果不包含请求参数 token 则不转发路由，否则执行正常的逻辑，代码如下所示。

```
@Component
public class AuthorizationFilter implements GlobalFilter, Ordered {
  @Override
  public Mono<Void> filter(ServerWebExchange exchange, GatewayFilterChain chain) {
    // 获取请求 URL
    String url = exchange.getRequest().getURI().getPath();
    // 忽略以下 URL 请求
    if (url.indexOf(/login) >= 0) {
        return chain.filter(exchange);
    }
    // 获取 token
    String token = exchange.getRequest().getQueryParams().getFirst(token);
    // 判断
    if (StringUtils.isBlank(token)) {
        System.out.println(token is empty ...);
        // 设置响应码
        exchange.getResponse().setStatusCode(HttpStatus.UNAUTHORIZED);
        // 响应
        return exchange.getResponse().setComplete();
    }
    // 如果后续有过滤器则进入下一个过滤器，否则正常响应
    return chain.filter(exchange);
  }
```

```
@Override
public int getOrder() {
    // 指定此过滤器的优先级，返回值越大级别越低
    return 0;
}
```

自定义全局过滤器需要实现 GlobalFilter 和 Ordered 接口。在 filter 方法中完成过滤器的逻辑判断处理，在 getOrder 方法中指定此过滤器的优先级，返回值越大级别越低。

ServerWebExchange 相当于当前请求和响应的上下文，存放着重要的请求响应属性、请求实例和响应实例等。一个请求中的 Request、Response 都可以通过 ServerWebExchange 调用 chain.filter 继续向下游执行。

重启服务后，访问 gateway-server，不包含 token 参数的访问结果如图 8-17 所示。

图8-17　不包含token参数的访问结果

重启服务后，访问 gateway-server，包含 token 参数的访问结果如图 8-18 所示。

图8-18　包含token参数的访问结果

8.5　网关限流

在高并发的应用中，限流是一个绕不开的话题。限流可以保障 API 服务对所有用户的可用性，也可以防止网络攻击。接下来一起研究一下 Gateway 中的网关限流。

8.5.1 常见的限流算法

常见的限流算法有计数器、漏桶算法和令牌桶算法 3 种，下面分别进行介绍。

■ 计数器

计数器限流算法是最简单的一种限流实现方法。其本质是通过维护一个单位时间内的计数器，每次请求时计数器加 1，当单位时间内计数器累加到大于设定的阈值时，之后的请求都会被拒绝，直到单位时间过去，再将计数器重置为 0。

■ 漏桶算法

漏桶算法可以很好地限制容量池的大小，从而防止流量暴增。漏桶可以看作一个带有常量服务时间的单服务器队列，如果漏桶（包缓存）溢出，那么数据包会被丢弃。在网络中，漏桶算法可以控制端口的流量输出速率，平滑网络上的突发流量，实现流量整形，从而为网络提供一个稳定的流量。

为了更好地控制流量，漏桶算法需要使用两个变量：一个是桶的大小，表示流量突发增多时可以存多少的"水"（Burst），另一个是桶漏洞的大小（Rate）。

■ 令牌桶算法

令牌桶算法是对漏桶算法的一种改进，漏桶算法能够限制请求调用的速率，而令牌桶算法能够在限制调用的平均速率的同时允许一定程度的突发调用。在令牌桶算法中，存在一个桶，用来存放固定数量的令牌。算法中存在一种机制，以一定的速率往桶中放令牌。每次请求调用需要先获取令牌，只有拿到令牌，才有机会继续执行，否则等待可用的令牌或者直接被拒绝。放令牌这个动作是持续不断地进行的，如果桶中令牌数达到上限，就丢弃令牌，所以存在这种情况，桶中一直有大量的可用令牌，这时进来的请求就可以直接拿到令牌。比如设置 QPS 为 100，那么限流器初始化完成 1s 后，桶中就已经有 100 个令牌了，这时服务还没完全启动好，等启动完成对外提供服务时，该限流器可以"抵挡"瞬时的 100 个请求。所以，只有桶中没有令牌时，请求才会进行等待，最后相当于以一定的速率调用请求。

8.5.2 基于过滤器的限流

Spring Cloud Gateway 官方提供了令牌桶的限流支持，是基于其内置的过滤器工厂 RequestRateLimiterGatewayFilterFactory 来实现的。在过滤器工厂中通过 Redis 和 Lua 脚本结合的方式进行流量控制。

■ 准备 Redis

下载 Redis 之后，解压到指定目录下，进入 Redis 目录后双击 redis-server.exe 启动 Redis 服务器，如图 8-19 所示。

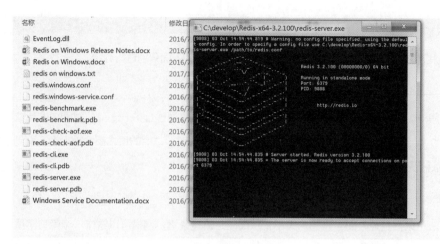

图8-19 启动Redis服务器

在图8-19所示页面中双击redis-cli.exe启动客户端，执行测试命令进行测试，如图8-20所示。

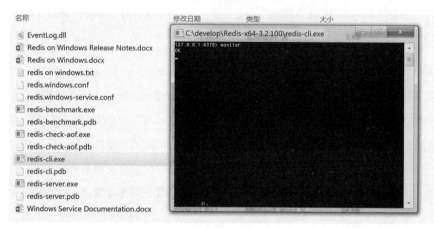

图8-20 启动Redis客户端

■ 引入 Redis 的依赖

在子工程 gateway-server 的 pom.xml 文件中引入 Redis 相关依赖，如下所示。

```xml
<!-- 监控依赖 -->
<dependency>
    <groupId>org.springframework.boot</groupId>
    <artifactId>spring-boot-starter-actuator</artifactId>
</dependency>
<!-- Redis 依赖 -->
<dependency>
    <groupId>org.springframework.boot</groupId>
    <artifactId>spring-boot-starter-data-redis-reactive</artifactId>
</dependency>
```

■ 修改 application.yml 配置文件

在 application.yml 配置文件中加入限流的配置，如下所示。

```yaml
server:
  port: 8080 # 服务端口
spring:
  application:
    name: gateway-server # 指定服务名称
  cloud:
    gateway:
      routes:
        - id: product-service
          uri: lb://product-service
          predicates:
            - Path=/product-service/**
          filters:
            - RewritePath=/product-service/(?<segment>.*),/$\{segment}
            - name: RequestRateLimiter
              args:
                key-resolver: #{@pathKeyResolver} # 使用 SpEL 从容器中获取对象
                redis-rate-limiter.replenishRate: 1 # 令牌桶每秒填充平均速率
                redis-rate-limiter.burstCapacity: 3 # 令牌桶的总容量
eureka: # 配置 Eureka
  client:
    service-url:
      defaultZone: http://10.211.55.12:9000/eureka/ # 多个 Eureka 服务端之间用逗号隔开
  instance:
    prefer-ip-address: true # 使用 IP 地址注册
    instance-id: ${spring.cloud.client.ip-address}:${server.port} # 向注册中心注册服务 id
```

在 application.yml 中添加了 Redis 的信息，并配置了 RequestRateLimiter，相关配置项如下。

（1）burstCapacity：令牌桶总容量。

（2）replenishRate：令牌桶每秒填充平均速率。

（3）key-resolver：用于限流的键的解析器的 Bean 对象的名字。它使用 SpEL 表达式根据 #{@beanName}# 从 Spring 容器中获取 Bean 对象。

■ 编写配置类

为了达到不同的限流效果和实现不同的限流规则，可以通过实现 KeyResolver 接口，定义不同请求类型的限流键，如下所示。

```java
@Configuration
public class KeyResolverConfiguration {

    /**
     * 编写基于请求路径的限流规则
     */
    @Bean
    public KeyResolver pathKeyResolver() {
        // 自定义的 KeyResolver
        return new KeyResolver() {
            /**
             * ServerWebExchange :
             *       上下文参数
             */
```

```
        public Mono<String> resolve(ServerWebExchange exchange) {
            return Mono.just(exchange.getRequest().getPath().toString());
        }
    };
}

/**
 * 编写基于请求 IP 地址的限流规则
 *
 * @return
 */
/*@Bean
public KeyResolver ipKeyResolver() {
    return exchange -> Mono.just(exchange.getRequest().getHeaders().getFirst(X-Forwarded-For));
}*/

/**
 * 编写基于请求参数的限流规则
 *
 * @return
 */
/*@Bean
public KeyResolver userKeyResolver() {
    return exchange -> Mono.just(exchange.getRequest().getQueryParams().getFirst(user));
}*/
}
```

■ 代码测试

使用 JMeter 模拟 5 组线程访问，会看到如图 8-21 所示的结果。当达到令牌桶的总容量时，其他的请求会返回 429 错误。

图8-21 基于过滤器的限流代码测试

通过 Reids 的 MONITOR 命令可以监听 Redis 的执行过程。这时 Redis 中会输出图 8-22 所示的数据信息。

图8-22 监听Redis的执行过程

图 8-22 的花括号中的就是限流的路径。timestamp 存储的是表示当前时间的秒数，也就是 System.currentTimeMillis() / 1000 或者 Instant.now().getEpochSecond() 的值。tokens 存储的是当前可用的令牌数量。Spring Cloud Gateway 目前提供的限流策略还是相对比较简单的，在实际应用中，我们会针对很多种情况调整限流策略，比如对不同接口进行限流和显示被限流后的友好提示信息等，这些可以通过自定义 RedisRateLimiter 来实现，这里不做讨论。

8.5.3 基于 Sentinel 的限流

Sentinel 支持对 Spring Cloud Gateway、Zuul 等主流的 API 网关进行限流，如图 8-23 所示。

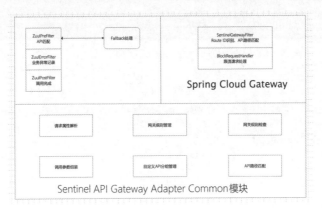

图8-23 Sentinel

从 1.6.0 版本开始，Sentinel 提供了 Spring Cloud Gateway 的适配模块，可以提供以下两种资源维度的限流。

（1）Route 维度：在 Spring 配置文件中配置的路由条目，资源名称为对应的路由 ID。

（2）自定义 API 维度：用户可以利用 Sentinel 提供的 API 来自定义 API 分组。

Sentinel 1.6.0 引入了 Sentinel API Gateway Adapter Common 模块，此模块中包含以下网关限流的规则和自定义 API 的实体和管理逻辑。

（1）GatewayFlowRule：网关限流规则，是针对 API 网关的场景定制的限流规则，可以对不同 Route 或自定义的 API 分组进行限流，支持对请求中的参数、Header、源 IP 地址等进行定制化的限流。

（2）ApiDefinition：用户自定义的 API 分组，可以看作一些 URL 匹配的组合。比如我们可以定义一个名为 my_api 的 API，请求路径模式为 /foo/** 和 /baz/** 的都归属到 my_api 这个 API 分组下面。限流的时候可以针对这个自定义的 API 分组维度进行。

■ 引入 Sentinel 的依赖

在子工程 gateway-server 的 pom.xml 文件中引入 Sentinel 相关依赖，如下所示。

```xml
<dependency>
    <groupId>com.alibaba.csp</groupId>
    <artifactId>sentinel-spring-cloud-gateway-adapter</artifactId>
    <version>1.6.3</version>
</dependency>
```

■ 修改 application.yml 配置文件

在 application.yml 配置文件中加入限流的配置，如下所示。

```yaml
server:
  port: 8080 # 服务端口
spring:
  application:
    name: gateway-server # 指定服务名称
  cloud:
    gateway:
      routes:
        - id: product-service
          uri: lb://product-service
          predicates:
            - Path=/product-service/**
          filters:
            - RewritePath=/product-service/(?<segment>.*),/$\{segment}
eureka: # 配置 Eureka
  client:
    service-url:
      defaultZone: http://10.211.55.12:9000/eureka/ # 多个 Eureka 服务端之间用逗号隔开
  instance:
    prefer-ip-address: true # 使用 IP 地址注册
    instance-id: ${spring.cloud.client.ip-address}:${server.port} # 向注册中心注册服务 id
```

■ 编写配置类

配置限流参数，指定限流规则。配置类的具体实现如下所示。

```java
@Configuration
public class GatewayConfiguration {

    private final List<ViewResolver> viewResolvers;

    private final ServerCodecConfigurer serverCodecConfigurer;

    public GatewayConfiguration(ObjectProvider<List<ViewResolver>> viewResolversProvider,
                ServerCodecConfigurer serverCodecConfigurer) {
        this.viewResolvers = viewResolversProvider.getIfAvailable(Collections::emptyList);
        this.serverCodecConfigurer = serverCodecConfigurer;
    }

    /**
     * 配置限流的异常处理器 :SentinelGatewayBlockExceptionHandler
     */
    @Bean
    @Order(Ordered.HIGHEST_PRECEDENCE)
    public SentinelGatewayBlockExceptionHandler sentinelGatewayBlockExceptionHandler() {
        return new SentinelGatewayBlockExceptionHandler(viewResolvers, serverCodecConfigurer);
    }

    /**
     * 配置限流过滤器
     */
    @Bean
    @Order(Ordered.HIGHEST_PRECEDENCE)
    public GlobalFilter sentinelGatewayFilter() {
        return new SentinelGatewayFilter();
    }

    /**
     * 配置初始化的限流参数
     * 用于指定资源的限流规则
     *     1. 资源名称（路由 ID）
     *     2. 配置统计时间
     *     3. 配置限流阈值
     */
    @PostConstruct
    public void initGatewayRules() {
        Set<GatewayFlowRule> rules = new HashSet<>();
rules.add(new GatewayFlowRule(product-service)
.setCount(3)
.setIntervalSec(1)
);
//      rules.add(new GatewayFlowRule(product_api)
//              .setCount(1).setIntervalSec(1)
//      );

        GatewayRuleManager.loadRules(rules);
```

```java
    }
    /**
     * 自定义 API 分组
     *   1. 定义分组
     *   2. 对分组配置限流规则
     */
    @PostConstruct
    private void initCustomizedApis() {
        Set<ApiDefinition> definitions = new HashSet<>();
        ApiDefinition api1 = new ApiDefinition(product_api)
            .setPredicateItems(new HashSet<ApiPredicateItem>() {{
                add(new ApiPathPredicateItem().setPattern(/product-service/product/**). // 以 /product-service/product/ 开头的所有 URL
                    setMatchStrategy(SentinelGatewayConstants.URL_MATCH_STRATEGY_PREFIX));
            }});
        ApiDefinition api2 = new ApiDefinition(order_api)
            .setPredicateItems(new HashSet<ApiPredicateItem>() {{
                add(new ApiPathPredicateItem().setPattern(/order-service/order)); // 完全匹配 /order-service/order 的 URL
            }});
        definitions.add(api1);
        definitions.add(api2);
        GatewayApiDefinitionManager.loadApiDefinitions(definitions);
    }

    /**
     * 自定义限流处理器
     */
    @PostConstruct
    public void initBlockHandlers() {
        BlockRequestHandler blockHandler = (serverWebExchange, throwable) -> {
            Map map = new HashMap();
            map.put(code,001);
            map.put(message, 不好意思，限流啦 );
            return ServerResponse.status(HttpStatus.OK)
                .contentType(MediaType.APPLICATION_JSON_UTF8)
                .body(BodyInserters.fromObject(map));
        };
        GatewayCallbackManager.setBlockHandler(blockHandler);
    }
}
```

基于 Sentinel 的网关限流是通过其提供的过滤器来完成的，使用时只需注入对应的 SentinelGatewayFilter 实例以及 SentinelGatewayBlockExceptionHandler 实例即可。

@PostConstruct 定义初始化的加载方法，用于指定资源的限流规则。上述代码中的资源名称为 product-service，统计时间为 1s，限流阈值为 3，表示每秒只能访问 3 个请求。

■ 代码测试

使用浏览器访问 product-service，当每秒访问超过 3 个请求时，结果如图 8-24 所示。

图8-24　Sentinel限流代码测试结果

8.6 源码解析

作为后端服务的统一入口，API 网关可提供请求路由、协议转换、安全认证、服务鉴权、流量控制与日志监控等服务。如果使用微服务架构将所有的应用管理起来，那么 API 网关就起到了微服务网关的作用。如果只是使用 REST 方式进行服务之间的访问，使用 API 网关对调用进行管理，那么 API 网关起到的就是服务治理的作用。不管哪一种使用方式，都不影响 API 网关核心功能的实现。当请求到达网关时，网关处理流程如图 8-25 所示。

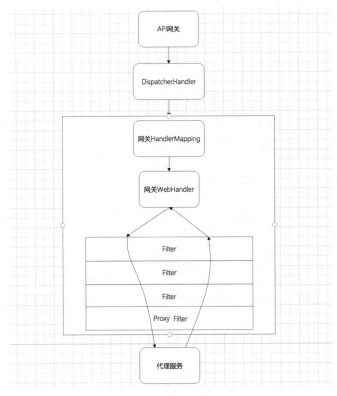

图8-25　网关处理流程

网关处理流程的具体步骤如下。

（1）请求发送到 API 网关，DispatcherHandler 是 HTTP 请求的中央分发处理器，负责将请求匹配到相应的 HandlerMapping。

（2）请求与处理器之间有一个映射关系，网关 Handler Mapping 会对请求进行路由匹配。

（3）请求随后到达网关的 Web 处理器，即 WebHandler，它代理了一系列网关过滤器和全局过滤器的实例，如对请求或者响应的头部进行处理（增加或者移除某个头部）。

（4）最后，请求转发给具体的代理服务。

这里比较重要的功能点是路由的过滤和路由的定位。Spring Cloud Gateway 提供了非常丰富的路由过滤器和路由断言。下面将按照自上而下的顺序分析网关处理流程部分的源码。

8.6.1 初始化配置

在引入 Spring Cloud Gateway 的依赖后，Starter 的 JAR 包将会自动初始化一些类，如下所示。

（1）GatewayLoadBalancerClientAutoConfiguration，客户端负载均衡配置类。

（2）GatewayRedisAutoConfiguration，Redis 的自动配置类。

（3）GatewayDiscoveryClientAutoConfiguration，服务发现自动配置类。

（4）GatewayClassPathWarning AutoConfiguration，WebFlux 依赖检查的配置类。

（5）GatewayAutoConfiguration，核心配置类，用于配置路由规则、过滤器等。

这些类的配置方式就不一一列出讲解了，主要看一下涉及的网关属性配置定义，很多对象的初始化都依赖于应用服务中配置的网关属性。GatewayProperties 是网关中主要的配置属性类，代码如下所示。

```
@ConfigurationProperties(spring.cloud.gateway)
@Validated
public class GatewayProperties {
    private final Log logger = LogFactory.getLog(this.getClass());
    // 路由列表
    @NotNull
    @Valid
    private List<RouteDefinition> routes = new ArrayList();
    private List<FilterDefinition> defaultFilters = new ArrayList();
    private List<MediaType> streamingMediaTypes;

    public GatewayProperties() {
        this.streamingMediaTypes = Arrays.asList(MediaType.TEXT_EVENT_STREAM, MediaType.APPLICATION_STREAM_JSON);
......
}
```

GatewayProperties 中有 3 个属性，分别是路由、默认过滤器和 MediaType。routes 是一个列表，对应的对象属性是路由定义 RouteDefinition。defaultFilters 是默认的路由过滤器，会应用到每个路由中。streamingMediaTypes 默认支持两种类型：APPLICATION_STREAM_JSON 和 TEXT_EVENT_STREAM。

8.6.2 网关处理器

请求到达网关之后，会有各种 Web 处理器对请求进行匹配与处理，图 8-26 所示为 Spring Cloud Gateway 中主要涉及的 WebHandler 核心类图。

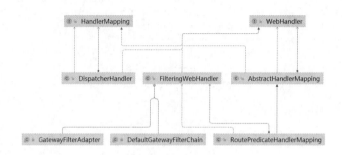

图8-26 WebHandler 核心类图

本小节将会按照以下顺序介绍负责请求路由选择和定位的处理器。

DispatcherHandler → RoutePredicateHandlerMapping → FilteringWebHandler → DefaultGatewayFilterChain。

■ 请求的分发处理器 DispatcherHandler

Spring Cloud Gateway 引入了 Spring WebFlux，DispatcherHandler 是其访问入口，是请求分发处理器。在之前的项目中，引入了 Spring MVC，而它的分发处理器是 DispatcherServlet。下面具体看一下网关收到请求后，如何匹配 HandlerMapping，代码如下所示。

```
public class DispatcherHandler implements WebHandler, ApplicationContextAware {
......
    public Mono<Void> handle(ServerWebExchange exchange) {
// 不存在 handlerMappings 则报错
        return this.handlerMappings == null ? this.createNotFoundError() : Flux.fromIterable(this.handlerMappings).concatMap((mapping) -> {
            return mapping.getHandler(exchange);
        }).next().switchIfEmpty(this.createNotFoundError()).flatMap((handler) -> {
            return this.invokeHandler(exchange, handler);
        }).flatMap((result) -> {
            return this.handleResult(exchange, result);
        });
    }
......
}
```

DispatcherHandler 实现了 WebHandler 接口，WebHandler 接口用于处理网络请求。DispatcherHandler 的构造方法会初始化 HandlerMapping，其核心处理的方法是 handle(ServerWebExchange exchange)，而 HandlerMapping 是一个定义了请求与处理器对象映射的接口且有多个实现类，如 ControllerEndpointHandlerMapping 和 RouterFunctionMapping，调试网关中的 handler 映射。

路由断言的 HandlerMapping

RoutePredicateHandlerMapping 用于匹配具体的路由，并返回处理路由的 FilteringWebHandler，代码如下所示。

```
public class RoutePredicateHandlerMapping extends AbstractHandlerMapping {
    public RoutePredicateHandlerMapping(FilteringWebHandler webHandler, RouteLocator routeLocator, GlobalCorsProperties globalCorsProperties, Environment environment) {
        this.webHandler = webHandler;
        this.routeLocator = routeLocator;
        if (environment.containsProperty(management.server.port)) {
            this.managmentPort = new Integer(environment.getProperty(management.server.port));
        } else {
            this.managmentPort = null;
        }

        this.setOrder(1);
        this.setCorsConfigurations(globalCorsProperties.getCorsConfigurations());
    }
    ......
}
```

RoutePredicateHandlerMapping 的构造方法接收两个参数：FilteringWebHandler（网关过滤器）和 RouteLocator（路由定位器）。setOrder(1) 用于设置该对象初始化的优先级。Spring Cloud Gateway 的 GatewayWebfluxEndpoint 提供的 HTTP API 不需要经过网关转发，它通过 RequestMappingHandlerMapping 进行请求匹配，因此需要将 RoutePredicateHandlerMapping 的优先级设置为低于 RequestMappingHandlerMapping 的优先级。

过滤器的 Web 处理器 Filtering Web Handler

FilteringWebHandler 通过创建所请求 route 对应的 GatewayFilterChain，在网关处进行过滤处理，代码如下所示。

```
public class FilteringWebHandler implements WebHandler {
......
    private static List<GatewayFilter> loadFilters(List<GlobalFilter> filters) {
        return (List)filters.stream().map((filter) -> {
// 适配器模式，用以适配 GlobalFilter
            FilteringWebHandler.GatewayFilterAdapter gatewayFilter = new FilteringWebHandler.GatewayFilterAdapter(filter);
// 判断是否实现 Ordered 接口
            if (filter instanceof Ordered) {
                int order = ((Ordered)filter).getOrder();
                return new OrderedGatewayFilter(gatewayFilter, order);
            } else {
                return gatewayFilter;
            }
        }).collect(Collectors.toList());
    }
......
    public Mono<Void> handle(ServerWebExchange exchange) {
```

```java
        Route route = (Route)exchange.getRequiredAttribute(ServerWebExchangeUtils.GATEWAY_
ROUTE_ATTR);
        List<GatewayFilter> gatewayFilters = route.getFilters();
        // 加入全局过滤器
 List<GatewayFilter> combined = new ArrayList(this.globalFilters);
        combined.addAll(gatewayFilters);
// 对过滤器排序
        AnnotationAwareOrderComparator.sort(combined);
        if (logger.isDebugEnabled()) {
          logger.debug(Sorted gatewayFilterFactories:  + combined);
        }
// 按照优先级，生成过滤器链，对该请求进行过滤
        return (new FilteringWebHandler.DefaultGatewayFilterChain(combined)).filter(exchange);
    }
......
    private static class DefaultGatewayFilterChain implements GatewayFilterChain {
......
      public Mono<Void> filter(ServerWebExchange exchange) {
        return Mono.defer(() -> {
          if (this.index < this.filters.size()) {
            GatewayFilter filter = (GatewayFilter)this.filters.get(this.index);
              FilteringWebHandler.DefaultGatewayFilterChain chain = new FilteringWebHandler.
DefaultGatewayFilterChain(this, this.index + 1);
            return filter.filter(exchange, chain);
          } else {
            return Mono.empty();
          }
        });
      }
    }
}
```

上述代码中，全局变量 globalFilters 是 Spring Cloud Gateway 中定义的全局过滤器，通过传入的全局过滤器，对这些过滤器进行适配处理。因为过滤器的定义有优先级，所以这里的处理主要是判断是否实现 Ordered 接口，如果实现了 Ordered 接口，则返回 OrderedGatewayFilter 对象，否则返回过滤器的适配器，用以适配 GlobalFilter。

FilteringWebHandler#handle 方法首先获取请求对应的路由的过滤器和全局过滤器，将两部分组合，然后对过滤器排序，AnnotationAwareOrderComparator 是 OrderComparator 的子类，支持对 Spring 的 Ordered 接口进行优先级排序，最后按照优先级，生成过滤器链，对该请求进行过滤处理。这里的过滤器链是通过内部静态类 DefaultGatewayFilterChain 实现的，该类实现了 GatewayFilterChain 接口，用于按优先级过滤。

8.6.3 路由定义定位器

RouteDefinitionLocator 是路由定义定位器的顶级接口，具体的路由定义定位器都继承自该接口。RouteDefinitionLocator 的类图如图 8-27 所示。

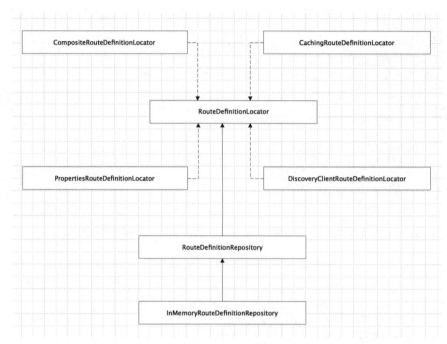

图8-27 RouteDefinitionLocator的类图

RouteDefinitionLocator 接口的定义如下所示。

```
public interface RouteDefinitionLocator {
    Flux<RouteDefinition> getRouteDefinitions();
}
```

接口中唯一的 getRouteDefinitions 方法用于获取路由定义。RouteDefinition 对象作为属性定义在 GatewayProperties 中，网关服务在启动时会读取配置文件中的相关配置。RouteDefinition 的定义如下所示。

```
@Validated
public class RouteDefinition {
    @NotEmpty
    private String id = UUID.randomUUID().toString();
    @NotEmpty
    @Valid
    private List<PredicateDefinition> predicates = new ArrayList();
    @Valid
    private List<FilterDefinition> filters = new ArrayList();
    @NotNull
    private URI uri;
    private int order = 0;
    ......
}
```

在 RouteDefinition 中，主要有 5 个属性，分别是 id（路由）、uri（转发地址）、order（优先级）、PredicateDefinition（路由断言定义）和 FilterDefinition（过滤器定义）。再深入的话，可以看到路由断言和过滤器属性是一个 Map 数据结构，用于存放多个对应的键值数组。通过 RouteDefinitionLocator 的类图，可以看出该接口有 4 个实现类，如下所示。

（1）基于属性配置的 PropertiesRouteDefinitionLocator。

（2）基于服务发现的 DiscoveryClientRouteDefinitionLocator。

（3）基于组合方式的 CompositeRouteDefinitionLocator。

（4）基于缓存方式的 CachingRouteDefinitionLocator。

在图 8-27 中，还有一个接口 RouteDefinitionRepository 继承自 RouteDefinitionLocator，用于对路由定义进行操作，如保存和删除路由定义。

8.6.4 路由定位器

直接获取路由的方法是通过路由定位器（RouteLocator）接口获取。同样，该顶级接口也有多个实现类，如图 8-28 所示为 RouteLocator 的类图。

图8-28　RouteLocator的类图

RouteLocator 接口的定义如下所示。

```
public interface RouteLocator {
    Flux<Route> getRoutes();
}
```

Route 定义了路由断言、过滤器、路由地址和路由的优先级等主要信息。请求到达后，会在转发到代理服务之前，依次经过路由断言进行匹配路由和网关过滤器处理。通过 RouteLocator 的类图，可以知道 RouteLocator 有 3 个实现类，如下所示。

（1）基于路由定义方式的 RouteDefinitionRouteLocator。

（2）基于缓存方式的 CachingRouteLocator。

（3）基于组合方式的 CompositeRouteLocator。

8.6.5 路由断言

Spring Cloud Gateway 创建 Route 对象时，使用 RoutePredicateFactory 创建 Predicate 对象。Predicate 对象可以赋值给 Route。简单来说，路由断言用于匹配请求对应的 Route。路由决策工厂 RoutePredicateFactory 的定义如下所示。

```
public interface RoutePredicateFactory<C> extends ShortcutConfigurable, Configurable<C> {
    String PATTERN_KEY = pattern;

    default Predicate<ServerWebExchange> apply(Consumer<C> consumer) {
        C config = this.newConfig();
        consumer.accept(config);
```

```
            this.beforeApply(config);
            return this.apply(config);
        }

        default AsyncPredicate<ServerWebExchange> applyAsync(Consumer<C> consumer) {
            C config = this.newConfig();
            consumer.accept(config);
            this.beforeApply(config);
            return this.applyAsync(config);
        }
......
}
```

RoutePredicateFactory 接口继承自 ShortcutConfigurable 接口，该接口将在后面的多个实现类中出现。基于传入的具体 RouteDefinitionLocator 获取路由定义时，用到了该接口中的默认方法。路由断言的种类很多，不同的路由断言需要的配置参数不一样，所以每种路由断言和过滤器的实现都会实现 ShortcutConfigurable 接口，用于指定自身参数个数和顺序。

8.6.6 网关过滤器

GatewayFilter（网关过滤器）用于拦截和链式处理 Web 请求，可以实现横切的、与应用无关的需求，比如提高安全性、访问超时的设定等。GatewayFilter 的定义如下所示。

```
public interface GatewayFilter extends ShortcutConfigurable {
    String NAME_KEY = name;
    String VALUE_KEY = value;

    Mono<Void> filter(ServerWebExchange exchange, GatewayFilterChain chain);
}
```

接口中定义了唯一的方法 filter，用于处理 Web 请求，并且可以通过给定的过滤器链将请求传递到下一个过滤器。该接口有多个实现类，其类图如图 8-29 所示。

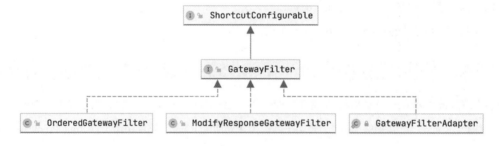

图8-29　GatewayFilter的类图

从图 8-29 中可以看出，GatewayFilter 有 3 个实现类，如下所示。

（1）ModifyResponseGatewayFilter 是一个内部类，用于修改响应体。

（2）OrderedGatewayFilter 是一个有序的网关过滤器。

（3）GatewayFilterAdapter 是一个适配器类，是定义在 Web 处理器中的内部类。

除此之外，GatewayFilterFactory 实现类的内部实际上创建了一个名为 GatewayFilter 的匿名类。

8.6.7 全局过滤器

GlobalFilter 接口与 GatewayFilter 具有相同的方法定义，该接口的设计和用法在将来的版本中可能会发生变化。GlobalFilter 是一系列特殊的过滤器，会根据条件应用到所有的路由中。之前已经了解了过滤器，定制的过滤器的粒度更细，由于有些过滤器需要全局应用，因此 Spring Cloud Gateway 中也提供了 GlobalFilter 的定义与实现。

下面通过类图来看一下 GlobalFilter 有哪些实现类，如图 8-30 所示。

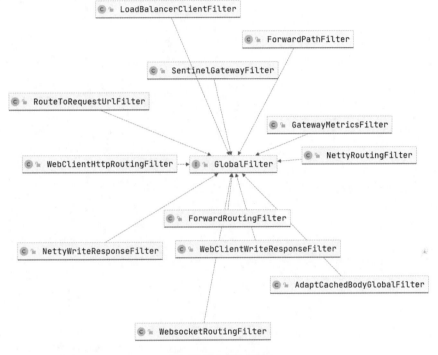

图8-30　GlobalFilter的类图

从图 8-30 中可以看到 GlobalFilter 有 12 个实现类，包括路由转发、负载均衡、WS 路由、Netty 路由等。

8.6.8　API 端点

Spring Cloud Gateway 提供了多个 API 端点，如过滤器列表、路由列表、单个路由信息等，用于路由相关的操作。Spring Cloud Gateway 的 API 端点纳管到 Spring Boot Actuator 中。在网关服务的启动日志中，可以看到网关的 API 端点。以下为 Spring Cloud Gateway 的主要 API 端点。

（1）/actuator/gateway/routes/ {id},methods=[DELETE]，用于删除单个路由。

（2）/actuator/gateway/routes/ {id},methods=[POST]，用于增加单个路由。

（3）/actuator/gateway/routes/ {id},methods=[GET]，用于查看单个路由。

（4）/actuator/gateway/routes,methods=[GET]，用于获取路由列表。

（5）/actuator/gateway/refresh,methods=[POST]，用于路由刷新。

（6）/actuator/gateway/globalfilters,methods=[GET]，用于获取全局过滤器列表。

（7）/actuator/gateway/routefilters,methods=[GET]，用于获取路由过滤器工厂列表。

（8）/actuator/gateway/routes/{id}/combinedfilters,methods=[GET]，用于获取单个路由的联合过滤器。

可以看到上述 API 端点包括 3 类：路由操作、获取过滤器和路由刷新。路由操作包括增加、删除、查看单个路由，以及获取路由列表等。获取过滤器有 3 个 API 端点：全局过滤器列表、过滤器工厂列表和单个路由的联合过滤器。

第 9 章

Spring Cloud Stream 消息驱动

在实际的企业微服务架构开发中，消息中间件是至关重要的组件之一。消息中间件主要解决应用解耦、异步消息和流量削锋等问题，以实现高性能、高可用、可伸缩和具备最终一致性的架构。不同的消息中间件其实现方式、内部结构是不一样的。如常见的 RabbitMQ 和 Kafka，由于这两个消息中间件架构上的不同，像 RabbitMQ 有 Exchange，Kafka 有主题和分区，因此在实际项目开发中会给使用者造成一定的困扰，如果用了两个消息中间件中的一个，后面的业务需求想往另外一个消息中间件迁移，无疑是灾难性的，一大堆东西都要重新推倒重新做，因为它和当前的系统耦合了。这个问题可以使用 Spring Cloud Stream 来解决，因为 Spring Cloud Stream 给使用者提供了一种解耦合的方法。

本章的主要内容如下。

1. 认识 Spring Cloud Stream。
2. 实现消息驱动。
3. 消费者组。
4. 消费分区。
5. 源码解析。

9.1 认识 Spring Cloud Stream

Spring Cloud Stream 是一个用来为微服务应用构建消息驱动能力的框架。微服务应用通过 Spring Cloud Stream 提供的输入（相当于消息消费者，它从队列中接收消息）和输出（相当于消息生产者，它向队列中发送消息）通道与外界交互。通道通过指定消息中间件的绑定层实现与外部代理的连接。通过 Spring Cloud Stream，业务开发者不再需要关注具体的消息中间件，只需关注 Binder 为微服务应用提供的抽象概念来使用消息中间件实现业务即可。Spring Cloud Stream 的架构如图 9-1 所示。

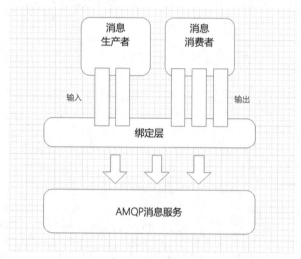

图9-1 Spring Cloud Stream的架构

在图 9-1 中，底层是 AMQP 消息服务，中间层是绑定层，绑定层和底层的 AMQP 消息服务绑定，顶层是消息生产者和消息消费者，顶层可以向绑定层生产消息和从绑定层获取消息来消费。

9.1.1 消息队列

消息中间件是分布式系统中十分重要的组件之一，主要用于解决应用耦合、异步消息和流量削锋等问题，是大型分布式系统不可缺少的中间件。消息队列技术是分布式应用间交换信息的一种技术，消息可驻留在内存或磁盘上，以队列形式存储消息，直到它们被应用程序消费。通过消息队列，应用程序可以相对独立地运行，它们不需要知道彼此的位置，只需要给消息队列发送消息并且处理从消息队列发送过来的消息。

消息队列的主要特点是异步处理和解耦。其主要的使用场景就是，将比较耗时且不需要同步返回结果的操作作为消息放入消息队列，从而实现异步处理；同时由于使用了消息队列，只要保证消息格式不变，消息的发送方和接收方并不需要直接联系，因此也不会受到对方的影响，这就是解耦。

下面以 RabbitMQ 消息队列为例，介绍消息队列的常用组件和相关概念。RabbitMQ 将整个消息队列应用分为下列几个组件：消息生产者（Producer）、交换器（Exchange）、队列（Queue）和消息消费者（Consumer）。

消息生产者就是生成消息并将消息发送给消息队列的应用，而消息消费者则是接收消息队列发送的消息并消费的应用。

图 9-2 所示是 RabbitMQ 内部用于存储消息的队列，每个队列都有唯一的名称（queue_name），用来让消息消费者进行订阅。消息生产者生产的消息会存储到队列中，消息消费者则从队列中获取并消费消息，三者的关系如图 9-3 所示。

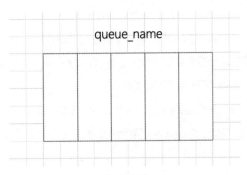

图9-2　队列

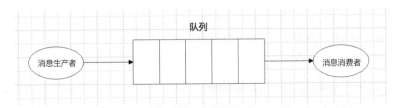

图9-3　消息生产者、队列、消息消费者三者的关系

多个消息消费者可以订阅同一个队列，这时队列中的消息会分配给多个消息消费者进行处理，而不是每一个消息消费者都收到所有的消息，如图 9-4 所示。

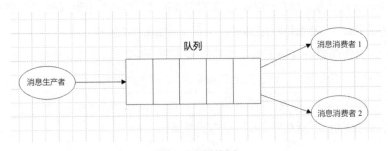

图9-4　多个消息消费者

虽然消息生产者可以直接将消息发送到队列中，但是在 RabbitMQ 中，消息生产者发送给消息队列的所有消息都要先经过交换器，再由交换器将消息路由到一个或者多个队列中。图 9-5 展示了交换器和队列之间的关系。

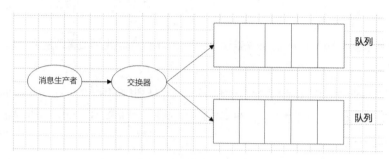

图9-5　交换器和队列之间的关系

在使用交换器时，需要将交换器和队列进行绑定。在绑定的同时，一般会指定一个绑定键值。而消息生产者在将消息发送给交换器时，一般会指定一个路由键值来指定这个消息的路由规则，而这个路由键值需要和交换器自身的类型以及绑定键值联合使用才能生效。

一般情况下当路由键值和绑定键值相匹配时，消息被路由到对应的队列中。但是绑定键值并不是所有情况都有效，这依赖于交换器的类型。RabbitMQ 常用的交换器类型有以下 4 种：Fanout、Direct、Topic 和 Headers。

（1）根据 Fanout 类型的交换器的路由规则，所有发送到该交换器的消息会被路由到所有与它绑定的队列中，可以将 Fanout 类型简单地理解为广播类型。

（2）根据 Direct 类型的交换器的路由规则，消息会被路由到绑定键值和该消息的路由键值完全匹配的队列中。

（3）Topic 类型的交换器在路由规则上与 Direct 类型的交换器类似，也是将消息路由到绑定键值与该消息路由键值相匹配的队列中，但它对路由规则进行了扩展，提供一些模糊路由规则。

（4）Headers 类型的交换器不依赖于路由键值与绑定键值的路由规则来路由消息，而是根据发送消息中的 headers 属性进行匹配。在绑定队列与交换器时会指定一组键值对。当消息发送到交换器时，RabbitMQ 会取到该消息的 headers（也是键值对的形式），对比其中的键值对是否完全匹配队列与交换器绑定时指定的键值对。如果完全匹配则消息会路由到该队列，否则不会路由到该队列。

RabbitMQ 发送消息的流程：首先获取网络连接，之后依次获取通道、定义交换器和队列。使用一个绑定键值将队列绑定到一个交换器上，通过指定交换器和路由键值将消息发送到对应的队列上，最后消息消费者在接收时也是先获取连接、通道。然后指定一个队列，从队列中获取消息。消息消费者对交换器、路由键值及如何绑定都不关心，只是从对应的队列中获取消息。

9.1.2　绑定器

绑定器（Binder）是 Spring Cloud Stream 中一个非常重要的概念。在没有绑定器这个概念的情况下，Spring Boot 应用在需要直接与消息中间件进行信息交互的时候，由于各消息中间件构建的初衷不同，它们的实现细节有较大的差异性，这使得实现的消息交互逻辑非常"笨重"，因为对具体的消息中间件实现细节有太重的依赖，当消息中间件有较大的变动、升级或更换消息中间件的时候，使用者就需要付出非常大的代价。

通过定义绑定器作为中间层实现了应用程序与消息中间件实现细节的隔离，通过向应用程序暴露统一的通道（Channel），使得应用程序不需要再考虑各种不同的消息中间件的实现。当需要升级消息中间件或者更换其他消息中间件产品时，使用者需要做的就是更换对应的绑定器而不是修改任何应用逻辑，使用者甚至可以任意地改变消息中间件的类型而不需要修改一行代码。

Spring Cloud Stream 支持各种绑定器，如下所示。

（1）RabbitMQ。

（2）Apache Kafka。

（3）Amazon Kinesis。

（4）Google PubSub。

（5）Solace PubSub。

（6）Azure Event Hubs。

通过配置把应用程序和 Spring Cloud Stream 的绑定器绑定在一起后，使用者只需要修改绑定器的配置来动态修改 Topic、Exchange 和 Type 等一系列信息，而不需要修改一行代码。

9.1.3 发布订阅模式

Spring Cloud Stream 中的消息通信方式遵循了发布订阅模式，当一条消息被投递到消息中间件之后，会通过共享的主题（Topic）进行广播，消息消费者在订阅的主题中收到它后触发自身的业务逻辑处理，如图 9-6 所示。

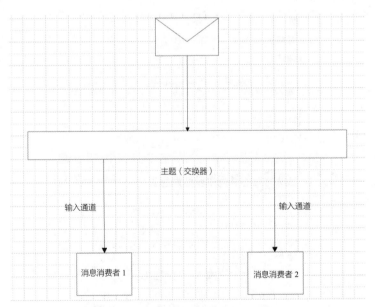

图9-6　发布订阅模式

这里所提到的主题是 Spring Cloud Stream 中的一个抽象概念，用来表示发布共享消息给消息消费者的地方。在不同的消息中间件中，主题可能对应着不同的概念，比如在 RabbitMQ 中它对应交换器，而在 Kafka 中则对应主题。

9.2 实现消息驱动

以 RabbitMQ 作为消息中间件，使用 Spring Cloud Stream 实现消息驱动，分别创建消息生产者和消息消费者子工程。

9.2.1 安装 RabbitMQ

RabbitMQ 服务端代码是使用并发式语言 Erlang 编写的，安装 RabbitMQ 的前提是安装 Erlang，分别访问 Erlang 和 RabbitMQ 官方网站进行下载，如图 9-7 所示。

图9-7 下载的软件

■ **安装 Erlang**

双击 Erlang 对应的 EXE 软件进行安装，安装成功之后，需要新建 Erlang 的环境变量 ERLANG_HOME，如图 9-8 所示。

图9-8 新建环境变量ERLANG_HOME

将环境变量 ERLANG_HOME 加入 Path 中,如图 9-9 所示。

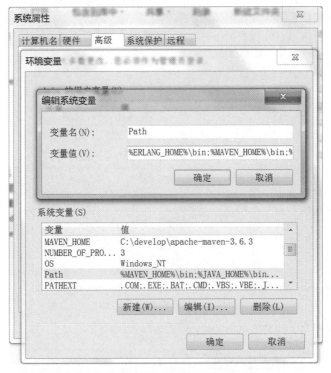

图9-9　将环境变量ERLANG_HOME加入Path中

打开命令行界面,执行 erl 命令,看到版本号就说明 Erlang 安装成功了,如图 9-10 所示。

图9-10　查看Erlang版本

安装 RabbitMQ

安装完成 Erlang,接下来安装 RabbitMQ。双击 RabbitMQ 对应的 EXE 软件进行安装,安装成功之后,需要新建 RabbitMQ 的环境变量 RABBITMQ_HOME,如图 9-11 所示。

图9-11 新建环境变量RABBITMQ_HOME

将环境变量 RABBITMQ_HOME 加入 Path，如图 9-12 所示。

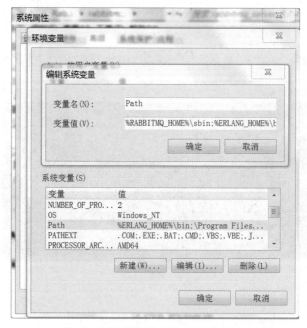

图9-12 将环境变量RABBITMQ_HOME加入Path

■ **开启插件**

rabbitmq_management 是管理后台的插件，要开启这个插件才能通过浏览器访问登录页面。在命令行界面中执行 rabbitmq-plugins enable rabbitmq_management 命令即可开启该插件，如图 9-13 所示。

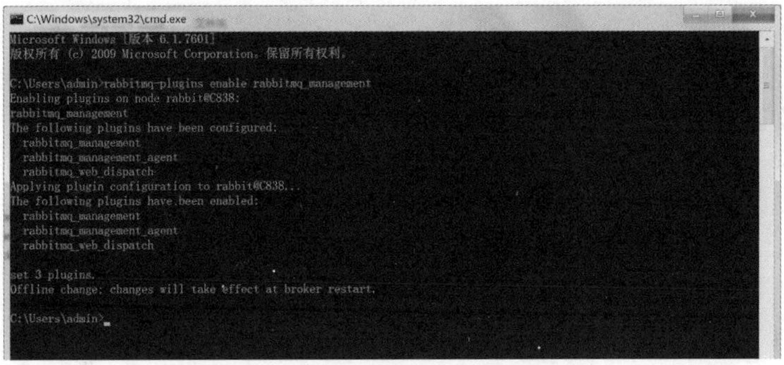

图9-13 开启插件

启动服务器

在命令行界面中执行 rabbitmq-server start 命令启动服务器，如图 9-14 所示。

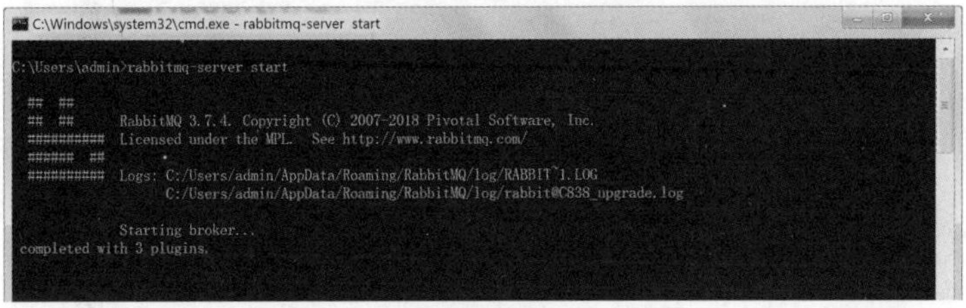

图9-14 启动服务器

登录管理后台

通过浏览器访问 http://localhost:15672，如图 9-15 所示。

图9-15 登录管理后台

使用默认的用户名 guest 和密码 guest 登录后，管理后台如图 9-16 所示。

图9-16 管理后台

9.2.2 消息生产者

使用 Spring Cloud Stream 实现发送消息到 RabbitMQ。

■ **创建子工程**

使用 IntelliJ IDEA 创建子工程——stream-producer，如图 9-17 所示。

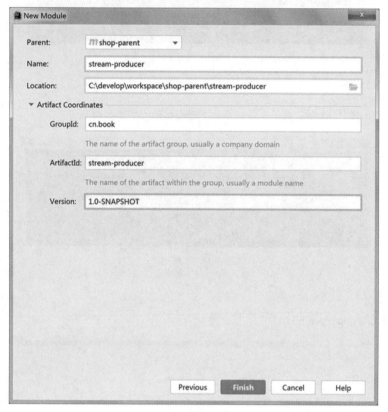

图9-17 创建stream-producer

■ **引入依赖**

在子工程 stream-producer 的 pom.xml 文件中引入相关依赖，如下所示。

```xml
<dependency>
    <groupId>org.springframework.cloud</groupId>
    <artifactId>spring-cloud-stream</artifactId>
</dependency>
<dependency>
    <groupId>org.springframework.cloud</groupId>
    <artifactId>spring-cloud-starter-stream-rabbit</artifactId>
</dependency>
<dependency>
    <groupId>org.springframework.cloud</groupId>
    <artifactId>spring-cloud-stream-binder-rabbit</artifactId>
</dependency>
```

■ 创建启动类

创建子工程 stream-producer 的启动类，如下所示。

```java
@SpringBootApplication
public class ProducerApplication {
    public static void main(String[] args) {
        SpringApplication.run(ProducerApplication.class);
    }
}
```

■ 创建配置文件

创建子工程 stream-producer 的配置文件 application.yml，如下所示。

```yaml
server:
  port: 6001
spring:
  application:
    name: stream-producer
  rabbitmq:
    addresses: 127.0.0.1
    username: guest
    password: guest
  cloud:
    stream:
      bindings:
        output:
          destination: book-default
      binders:
        defaultRabbit:
          type: rabbit
```

destination 指定了消息发送的目的地，对应 RabbitMQ 会发送消息到交换器 book-default，type 指定了使用的消息中间件是 RabbitMQ。

■ 声明和绑定通道

使用 Source 声明和绑定输出通道，并实现发送消息的方法，如下所示。

```java
@Component
@EnableBinding(Source.class)
public class MessageSender {

    @Autowired
    private MessageChannel output;

    // 发送消息
    public void send(Object obj) {
        output.send(MessageBuilder.withPayload(obj).build());
    }
}
```

代码测试

使用 @SpringBootTest 实现代码测试，如下所示。

```
@RunWith(SpringJUnit4ClassRunner.class)
@SpringBootTest
public class ProducerTest {
  @Autowired
  private MessageSender messageSender;

  @Test
  public void testSend() {
    for (int i = 1; i < 11; i++) {
      messageSender.send("msg" + i);
    }
  }
}
```

运行测试方法，发送消息，再次访问管理后台，查看交换器，如图 9-18 所示。

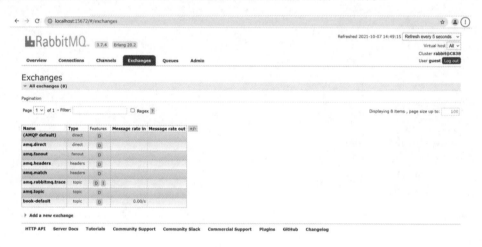

图9-18　通过管理后台查看Exchanges

在图 9-18 中查看交换器，发现已经创建了一个新的交换器，名称为 book-default，与 application.yml 中的 destination 的值一致。

9.2.3　消息消费者

使用 Spring Cloud Stream 实现从 RabbitMQ 中读取消息。

创建子工程

使用 IntelliJ IDEA 创建子工程——stream-consumer，如图 9-19 所示。

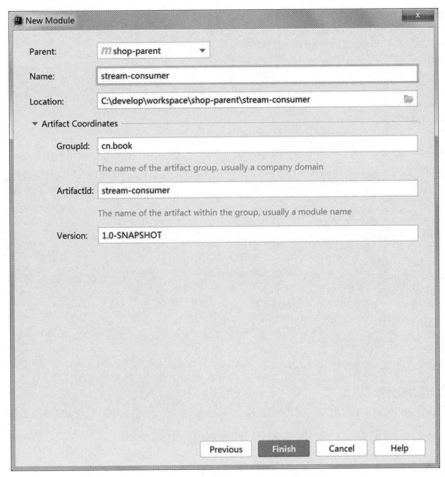

图9-19 创建stream-consumer

■ 引入依赖

在子工程 stream-consumer 的 pom.xml 文件中引入相关依赖,如下所示。

```
<dependency>
   <groupId>org.springframework.cloud</groupId>
   <artifactId>spring-cloud-stream</artifactId>
</dependency>
<dependency>
   <groupId>org.springframework.cloud</groupId>
   <artifactId>spring-cloud-starter-stream-rabbit</artifactId>
</dependency>
<dependency>
   <groupId>org.springframework.cloud</groupId>
   <artifactId>spring-cloud-stream-binder-rabbit</artifactId>
</dependency>
```

■ 创建启动类

创建子工程 stream-consumer 的启动类,如下所示。

```
@SpringBootApplication
public class ConsumerApplication {
    public static void main(String[] args) {
        SpringApplication.run(ConsumerApplication.class);
    }
}
```

■ 创建配置文件

创建子工程 stream-consumer 的配置文件 application.yml，如下所示。

```
server:
  port: 6002
spring:
  application:
    name: stream-consumer
  rabbitmq:
    addresses: 127.0.0.1
    username: guest
    password: guest
  cloud:
    stream:
      bindings:
        input:
          destination: book-default
      binders:
        defaultRabbit:
          type: rabbit
```

这里的 destination 的值与 stream-producer 中 destination 的值保持一致。

■ 声明和绑定通道

使用 Sink 声明和绑定输出通道，并实现读取消息的方法，如下所示。

```
@Component
@EnableBinding(Sink.class)
public class MessageReceiver {

    // 监听绑定中的消息
    @StreamListener(Sink.INPUT)
    public void input(String message) {
        System.out.println(" 获取到消息 : " + message);
    }
}
```

■ 代码测试

启动子工程 stream-consumer，监听等待消息，再次运行 stream-producer 发送消息的测试方法后，查看 stream-consumer 的控制台输出的日志，可查看读取的消息，如下所示。

```
........
获取到消息 : msg1
获取到消息 : msg2
```

```
获取到消息：msg3
获取到消息：msg4
获取到消息：msg5
获取到消息：msg6
获取到消息：msg7
获取到消息：msg8
获取到消息：msg9
获取到消息：msg10
……
```

9.2.4 自定义消息通道

Spring Cloud Stream 内置了两种接口，分别定义了绑定为"input"的输入流和"output"的输出流，而在实际使用中，往往需要定义各种输入输出流。接下来实现自定义消息通道。

■ **创建通道接口**

创建一个与 Source 和 Sink 类似的接口，代码如下所示。

```
/**
 * 自定义的消息通道
 */
public interface MyProcessor {

    /**
     * 消息生产者的配置
     */
    String MYOUTPUT = "myoutput";

    @Output(MYOUTPUT)
    MessageChannel myoutput();

    /**
     * 消息消费者的配置
     */
    String MYINPUT = "myinput";

    @Input(MYINPUT)
    SubscribableChannel myinput();
}
```

在一个项目中，可以定义无数个输入输出流，并可以根据实际业务情况进行划分。上述接口定义了一个输入、一个输出和两个绑定。

使用时，需要在 @EnableBinding 注解中添加自定义的接口。使用 @StreamListener 监听的时候，需要指定 MyProcessor.MYINPUT。

■ **工程改造**

修改子工程 stream-producer 中的 MessageSender，如下所示。

```
@Component
```

```java
@EnableBinding(MyProcessor.class)
public class MessageSender {

    @Autowired
    private MessageChannel myoutput;

    // 发送消息
    public void send(Object obj) {
        myoutput.send(MessageBuilder.withPayload(obj).build());
    }
}
```

修改子工程 stream-producer 的配置文件 application.yml，如下所示。

```yaml
server:
  port: 6001
spring:
  application:
    name: stream-producer
  rabbitmq:
    addresses: 127.0.0.1
    username: guest
    password: guest
  cloud:
    stream:
      bindings:
        myoutput::
          destination: my-book-default
      binders:
        defaultRabbit:
          type: rabbit
```

修改子工程 stream-consumer 中的 MessageReceiver，如下所示。

```java
@Component
@EnableBinding(MyProcessor.class)
public class MessageReceiver {

    // 监听绑定中的消息
    @StreamListener(MyProcessor.MYINPUT)
    public void input(String message) {
        System.out.println("获取到消息：" + message);
    }
}
```

修改子工程 stream-consumer 的配置文件 application.yml，如下所示。

```yaml
server:
  port: 6002
spring:
  application:
    name: stream-consumer
  rabbitmq:
    addresses: 127.0.0.1
    username: guest
```

```
      password: guest
  cloud:
    stream:
      bindings:
        myinput:
          destination: my-book-default
      binders:
        defaultRabbit:
          type: rabbit
```

■ **代码测试**

启动子工程 stream-consumer，监听等待消息，再次运行 stream-producer 发送消息的测试方法后，查看 stream-consumer 的控制台输出的日志，可查看读取的消息，如下所示。

```
……
  获取到消息 : msg1
  获取到消息 : msg2
  获取到消息 : msg3
  获取到消息 : msg4
  获取到消息 : msg5
  获取到消息 : msg6
  获取到消息 : msg7
  获取到消息 : msg8
  获取到消息 : msg9
  获取到消息 : msg10
……
```

9.3 消费者组

虽然 Spring Cloud Stream 通过发布订阅模式将消息生产者与消息消费者进行了很好的解耦，基于相同主题的消息消费者可以轻松地进行扩展，但是这些扩展都是针对不同的应用实例的。在现实的微服务架构中，每一个微服务应用为了实现高可用和负载均衡，实际上都会部署多个实例。在很多情况下，消息生产者发送消息给某个具体的微服务时，只希望消息被消费一次。为了解决这个问题，Spring Cloud Stream 提供了消费者组的概念。

如果同一个主题上的应用需要启动多个实例，可以通过 spring.cloud.stream.bindings.input.group 属性为应用指定一个组名，这样这个应用的多个实例在接收到消息的时候，只会有一个成员真正收到消息并进行消费。如图 9-20 所示，为 Service-A 和 Service-B 分别启动了两个实例，并且根据服务名称进行了分组，这样当消息进入主题之后，Group-A 和 Group-B 都会收到消息的副本，但是在两个组中都只会有一个实例对其进行消费。

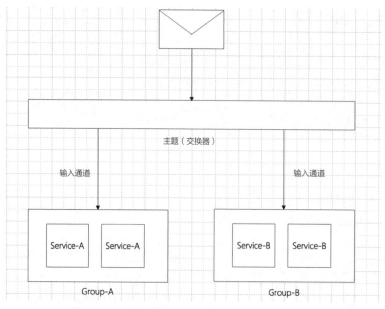

图9-20 消费者组

默认情况下,当没有为应用指定消费者组的时候,Spring Cloud Stream 会为其分配一个独立的匿名消费者组。所以,如果同一主题下的所有应用都没有被指定消费者组,当有消息发布之后,所有的应用都会对其进行消费,因为它们各自属于一个独立的组。大部分情况下,在创建 Spring Cloud Stream 应用的时候,建议为其指定一个消费者组,以防止对消息的重复处理,除非该行为需要这样做(比如刷新所有实例的配置等)。

9.3.1 工程改造

以9.2.2小节的消息生产者为例,创建两个消息生产者,端口号分别是6002和6003。消息消费者使用9.2.3小节的消息消费者。目前的项目结构如图 9-21 所示。

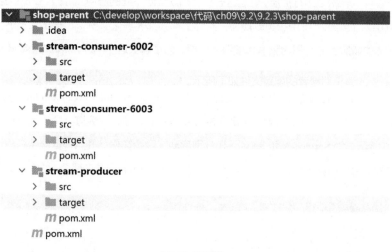

图9-21 项目结构

消息生产者的配置文件如下所示。

```yaml
server:
  port: 6001
spring:
  application:
    name: stream-producer
  rabbitmq:
    addresses: 127.0.0.1
    username: guest
    password: guest
  cloud:
    stream:
      bindings:
        output:
          destination: book-default-group
      binders:
        defaultRabbit:
          type: rabbit
```

destination 指定了消息发送的目的地，与消息消费者中的 destination 的值保持一致。

端口号为 6002 的消息消费者的配置文件如下所示。

```yaml
server:
  port: 6002
spring:
  application:
    name: stream-consumer
  rabbitmq:
    addresses: 127.0.0.1
    username: guest
    password: guest
  cloud:
    stream:
      bindings:
        input:
          destination: book-default-group
          group: mygroup
      binders:
        defaultRabbit:
          type: rabbit
```

端口号为 6003 的消息消费者的配置文件如下所示。

```yaml
server:
  port: 6003
spring:
  application:
    name: stream-consumer
  rabbitmq:
    addresses: 127.0.0.1
    username: guest
    password: guest
  cloud:
    stream:
```

```
      bindings:
        input:
          destination: book-default-group
          group: mygroup
      binders:
        defaultRabbit:
          type: rabbit
```

两个消息消费者中设置的 group 的值是一致的，表示二者属于同一个消费者组，一起完成消息的消费，不会重复。

9.3.2 代码测试

分别运行两个消息消费者和一个消息生产者，端口号为 6002 的消息消费者的控制台输出的日志如下所示。

```
获取到消息 : msg1
获取到消息 : msg3
获取到消息 : msg6
获取到消息 : msg7
获取到消息 : msg9
```

端口号为 6003 的消息消费者的控制台输出的日志如下所示。

```
获取到消息 : msg2
获取到消息 : msg4
获取到消息 : msg5
获取到消息 : msg8
获取到消息 : msg10
```

通过输出的日志可以发现，消费者组中的多个消息消费者共同消费消息。

9.4 消费分区

通过引入消费者组的概念，已经能够在有多个实例的情况下，保障每个消息只被组内的一个实例消费。通过上面设置消费者组参数的实验可以观察到，消费者组无法控制消息具体被哪个实例消费。也就是说，对于同一条消息，它多次到达之后可能是由不同的实例进行消费的。但是对于一些业务场景，需要将一些具有相同特征的消息设置为每次都被同一个实例消费。比如，一些监控服务，为了统计某段时间内消息生产者发送的报告内容，监控服务需要聚合这些数据，那么消息生产者可以为消息增加一个固有的特征 ID 来进行分区，使得拥有这些特征 ID 的消息每次都能被发送到一个特定的实例上以实现累计统计，否则这些数据就会分散到各个不同的实例，从而导致监控结果不一致。而分区（Partition）概念的引入就是为了解决这样的问题。当消息生产者将消息发送给多个实例时，需要保证拥有共同特征 ID 的消息始终是由同一个实例接收和处理的，如图 9-22 所示。

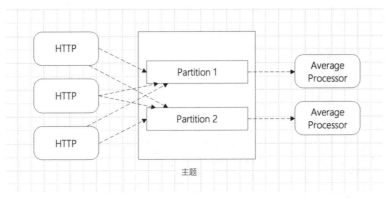

图9-22 分区

Spring Cloud Stream 为分区提供了通用的抽象实现,用来在消息中间件的上层实现分区处理,所以它对消息中间件自身是否实现了分区并不关心,这使得 Spring Cloud Stream 为不具备分区功能的消息中间件实现了分区功能扩展。

9.4.1 工程改造

以 9.3 节中的代码为基础进行工程改造。

消息生产者

创建消息实体类,包含消息 ID 和消息主体,如下所示。

```
@Data
@AllArgsConstructor
@NoArgsConstructor
public class Message implements Serializable {
    // 消息 ID
    private int id;
    // 消息主体
    private String body;
}
```

这里使用了 Lombok,所以在 pom.xml 文件中需要添加对应的依赖,如下所示。

```
<dependency>
    <groupId>org.projectlombok</groupId>
    <artifactId>lombok</artifactId>
</dependency>
```

修改配置文件 application.yml,如下所示。

```
server:
  port: 6001
spring:
  application:
    name: stream-producer
  rabbitmq:
    addresses: 127.0.0.1
```

```yaml
      username: guest
      password: guest
    cloud:
      stream:
        bindings:
          output:
            destination: book-default-group
            producer:
              partition-key-expression: payload.id
              partition-count: 2
        binders:
          defaultRabbit:
            type: rabbit
```

从上面的配置中，可以看到增加了以下两个参数。

（1）spring.cloud.stream.bindings.output.producer.partition-key-expression：通过该参数指定分区键的表达式规则，可以根据实际的输出消息规则来配置 SpEL 以生成合适的分区键。使用 payload.id 表示以消息对象的 id 属性作为分区键。

（2）spring.cloud.stream.bindings.output.producer.partition-count：通过该参数指定分区数。

■ 消息消费者

端口号为 6002 的消息消费者的配置文件如下所示。

```yaml
server:
  port: 6002
spring:
  application:
    name: stream-consumer
  rabbitmq:
    addresses: 127.0.0.1
    username: guest
    password: guest
  cloud:
    stream:
      instanceCount: 2
      instanceIndex: 0
      bindings:
        input:
          destination: book-default-group
          group: mygroup
          consumer:
            partitioned: true
      binders:
        defaultRabbit:
          type: rabbit
```

从上面的配置中，可以看到增加了以下 3 个参数。

（1）spring.cloud.stream.bindings.input.consumer.partitioned：通过该参数开启分区功能。

（2）spring.cloud.stream.instanceCount：通过该参数指定当前消息消费者的总实例数。

（3）spring.cloud.stream.instanceIndex：通过该参数设置当前实例的索引值，从 0 开始，最大值为 spring.cloud.stream.instanceCount-1。实验的时候需要启动多个实例，可以通过运行参数为不同实例设置不同的索引值。

端口号为 6003 的消息消费者的配置文件如下所示。

```yaml
server:
  port: 6003
spring:
  application:
    name: stream-consumer
  rabbitmq:
    addresses: 127.0.0.1
    username: guest
    password: guest
  cloud:
    stream:
      instanceCount: 2
      instanceIndex: 1
      bindings:
        input:
          destination: book-default-group
          group: mygroup
          consumer:
            partitioned: true
      binders:
        defaultRabbit:
          type: rabbit
```

注意，上面配置中的 instanceIndex 修改成 1 了。

9.4.2 代码测试

修改消息生产者的测试代码，改为发送消息对象，如下所示。

```java
@RunWith(SpringJUnit4ClassRunner.class)
@SpringBootTest
public class ProducerTest {
    @Autowired
    private MessageSender messageSender;

    @Test
    public void testSend() {
        for (int i = 1; i < 11; i++) {
            Message message1 = new Message(1,"msg"+i);
            messageSender.send(message1);

            Message message2 = new Message(2,"msg"+i);
            messageSender.send(message2);
        }
    }
}
```

上面的代码发送了 10 条 ID 为 1 的消息、10 条 ID 为 2 的消息。

分别运行两个消费者和一个生产者，端口号为 6002 的消息消费者的控制台输出的日志如下所示。

```
获取到消息：{"id":2,"body":"msg1"}
获取到消息：{"id":2,"body":"msg2"}
获取到消息：{"id":2,"body":"msg3"}
获取到消息：{"id":2,"body":"msg4"}
获取到消息：{"id":2,"body":"msg5"}
获取到消息：{"id":2,"body":"msg6"}
获取到消息：{"id":2,"body":"msg7"}
获取到消息：{"id":2,"body":"msg8"}
获取到消息：{"id":2,"body":"msg9"}
获取到消息：{"id":2,"body":"msg10"}
```

端口号为 6003 的消息消费者的控制台输出的日志如下所示。

```
获取到消息：{"id":1,"body":"msg1"}
获取到消息：{"id":1,"body":"msg2"}
获取到消息：{"id":1,"body":"msg3"}
获取到消息：{"id":1,"body":"msg4"}
获取到消息：{"id":1,"body":"msg5"}
获取到消息：{"id":1,"body":"msg6"}
获取到消息：{"id":1,"body":"msg7"}
获取到消息：{"id":1,"body":"msg8"}
获取到消息：{"id":1,"body":"msg9"}
获取到消息：{"id":1,"body":"msg10"}
```

通过输出的日志可以发现具有相同 ID 的消息每次都能发送到一个特定的实例上。原因是 partition-key-expression 设置为 payload.id，实例 payload 表示定义的消息，Spring Cloud Stream 默认在获取到这个消息后，通过 Hash 的方法进行计算，计算公式如下所示。

```
key.hashCode() % partitionCount
```

上面的 key 为消息 ID。ID 为 1 的消息根据计算公式得到 1，与端口号为 6003 的消息消费者的 instanceIndex 的值一致，所以该消息被分配给了端口号为 6003 的消息消费者。同理 ID 为 2 的消息被分配给了端口号为 6002 的消息消费者。

9.5 源码解析

Spring Cloud Stream 首先会动态注册相关 BeanDefinition，并且处理 @StreamListener 注解。然后会在 Bean 实例初始化之后，调用 BindingService 的相关方法进行服务绑定。BindingService 在绑定服务时会首先获取特定的绑定器，然后绑定消息生产者和消息消费者。最后 Spring Cloud Stream 的相关实例才会执行发送和接收消息的操作。Spring Cloud Stream 源码流程如图 9-23 所示。

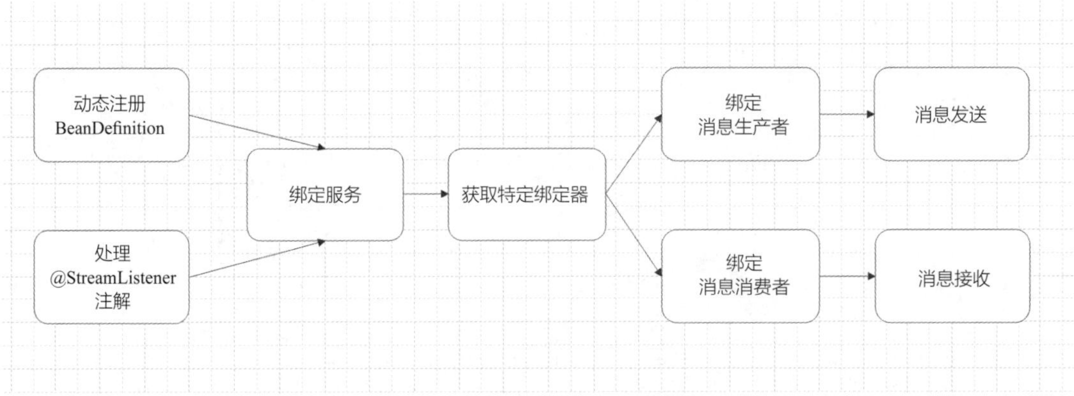

图9-23 Spring Cloud Stream源码流程

9.5.1 动态注册 BeanDefinition

@EnableBinding注解是Spring Cloud Stream生效的起点，它会将消息通道绑定到自己修饰的目标实例上，从而让这些实例具备与消息队列进行消息交互的能力。

@EnableBinding 使用 @Import 注解导入了 3 个类，分别是 BindingServiceConfiguration、BindingBeansRegistrar 和 BinderFactoryConfiguration。@EnableBinding 注解的具体实现如下所示。

```
...
@Import({BindingBeansRegistrar.class, BinderFactoryConfiguration.class})
@EnableIntegration
public @interface EnableBinding {
    Class<?>[] value() default {};
}
```

BindingBeansRegistrar 实现了 ImportBeanDefinitionRegistrar 接口，该接口用于向 Spring 容器动态注册 BeanDefinition。而 BindingBeansRegistrar 的作用是注册声明通道的接口类的 BeanDefinition，从而获取这些接口类的实例，并使用这些实例进行消息的发送和接收。

registerBeanDefinitions 方法需要动态注册 @EnableBinding 注解声明的接口类的 BeanDefinition，比如基础应用中的 MessageOutput。该方法会遍历 @EnableBinding 的 value 属性值的 Class 数组，然后依次调用 BindingBeanDefinitionRegistryUtils 的 registerBindingTargetBeanDefinitions 方法和 registerBindingTargetsQualified BeanDefinitions 方法进行 BeanDefinition 的动态注册，如下所示。

```
public class BindingBeansRegistrar implements ImportBeanDefinitionRegistrar {
......
    public void registerBeanDefinitions(AnnotationMetadata metadata, BeanDefinitionRegistry registry) {
......
//collectClasses 获取 @EnableBinding 注解的 value 属性值
// 因为这些接口的实例需要自动装配，所以这里必须给出这些实例的定义 BeanDefinition
//getClassName 获得的类信息是使用 @EnableBinding 注解的类信息
        for(int var6 = 0; var6 < var5; ++var6) {
```

```java
            Class<?> type = var4[var6];
            if (!registry.containsBeanDefinition(type.getName())) {
                BindingBeanDefinitionRegistryUtils.registerBindingTargetBeanDefinitions(type, type.getName(), registry);
                BindingBeanDefinitionRegistryUtils.registerBindingTargetsQualifiedBeanDefinitions(ClassUtils.resolveClassName(metadata.getClassName(), (ClassLoader)null), type, registry);
            }
        }
    }

    private Class<?>[] collectClasses(AnnotationAttributes attrs, String className) {
        EnableBinding enableBinding = (EnableBinding)AnnotationUtils.synthesizeAnnotation(attrs, EnableBinding.class, ClassUtils.resolveClassName(className, (ClassLoader)null));
// 通过 @EnableBinding 注解获取其值
        return enableBinding.value();
    }
}
```

registerBindingTargetBeanDefinition 方法会调用 ReflectionUtils 的 doWithMethods 方法来扫描接口类中的所有方法，然后为被 @Inpput 和 @Output 注解修饰的函数注册其返回值 SubscribableChannel 或者 MessageChannel 的 BeanDefinition，如下所示。

```java
public abstract class BindingBeanDefinitionRegistryUtils {
......
    private static void registerBindingTargetBeanDefinition(Class<? extends Annotation> qualifier, String qualifierValue, String name, String bindingTargetInterfaceBeanName, String bindingTargetInterfaceMethodName, BeanDefinitionRegistry registry) {
        if (registry.containsBeanDefinition(name)) {
            throw new BeanDefinitionStoreException(bindingTargetInterfaceBeanName, name, "bean definition with this name already exists - " + registry.getBeanDefinition(name));
        } else {
            RootBeanDefinition rootBeanDefinition = new RootBeanDefinition();
// 该 Bean 的实例由下面注册的 FactoryBean 和相应的 factorymethod 生成
            rootBeanDefinition.setFactoryBeanName(bindingTargetInterfaceBeanName);
            rootBeanDefinition.setUniqueFactoryMethodName(bindingTargetInterfaceMethodName);
            rootBeanDefinition.addQualifier(new AutowireCandidateQualifier(qualifier, qualifierValue));
            registry.registerBeanDefinition(name, rootBeanDefinition);
        }
    }

    public static void registerBindingTargetBeanDefinitions(Class<?> type, final String bindingTargetInterfaceBeanName, final BeanDefinitionRegistry registry) {
// 该方法通过 ReflectionUtils 来处理 MessageOutput 和 MessageInput 类中的被 @Input 和 @Output
注解修饰的函数
// 因为这些函数需要返回 MessageChannel 或者 SubscribableChannel 实例
        ReflectionUtils.doWithMethods(type, (method) -> {
            Input input = (Input)AnnotationUtils.findAnnotation(method, Input.class);
            if (input != null) {
                String name = getBindingTargetName(input, method);
                registerInputBindingTargetBeanDefinition(input.value(), name, bindingTargetInterfaceBeanName, bindingTargetInterfaceBeanName, method.getName(), registry);
            }
```

```
        Output output = (Output)AnnotationUtils.findAnnotation(method, Output.class);
        if (output != null) {
            String namex = getBindingTargetName(output, method);
                        registerOutputBindingTargetBeanDefinition(output.value(), namex, binding
TargetInterfaceBeanName, method.getName(), registry);
        }

    });
    }
    ......
}
```

使用 RootBeanDefinition 的 setFactoryBeanName 和 setUniqueFactoryMethodName 方法，将该 Bean 的实例化交给工厂类的对应函数来完成。

bindingTargetInterfaceBeanName 的值是声明通道的接口类的名称，而 bindingTargetInterfaceMethodName 的值对应函数的名称。Spring Cloud Stream 会为每个声明通道的接口类注册一个对应的工厂类，并且每个接口类中的函数都会有一个对应的工厂函数，所以这些函数的返回对象由对应的工厂函数生成。

registerBindingTargetsQualifiedBeanDefinitions 是在注册 registerBindingTargetBeanDefinition 时使用到的工厂类 BeanDefinition，这个工厂类用来生成 registerBindingTargetBeanDefinition 注册的 Bean 的实例。

Spring Cloud Stream 将 BindableProxyFactory 注册成了名称为 com.example.demo.message.MessageOutput 的 FactoryBean。通过 BindableProxyFactory，Spring 容器可以取到 registerBindingTargetBeanDefinitions 方法中所注册的 Bean 实例，例如 MessageOutput 中 outputMessagefunction 方法返回的 MessageChannel 实例。

9.5.2 消息发送的流程

Spring Cloud Stream 通过 MessageChannel 发送消息，MessageChannel 首先将消息发送通过 doBindProducer 绑定的 SendingHandler 对象，然后交由 MessageHandler 完成发送前的准备工作，最后由 RabbitTemplate 调用 RabbitMQ 客户端的 Channel 实例将消息发送给 RabbitMQ 服务端，如图 9-24 所示。

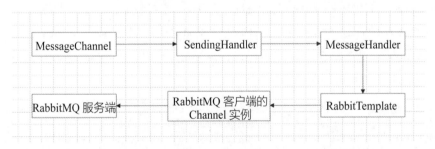

图9-24　消息发送流程

SendingHandler 是 AbstractMessageHandler 的子类，MessageChannel 有消息时，会将消息分发给 SendingHandler 的 handleMessageInternal 方法，然后由 SendingHandler 转发给对应的 MessageHandler。handleMessageInternal 方法的具体实现如下所示。

```
// SendingHandler.java
protected void handleMessageInternal(org.springframework.messaging.Message<?> message)
throws Exception {
    org.springframework.messaging.Message<?> messageToSend = this.useNativeEncoding ?
message : this.serializeAndEmbedHeadersIfApplicable(message);
    this.delegate.handleMessage(messageToSend);
}
```

delegate 是 createProducerMessageHandler 方法生成的 MessageHandler 实例,也就是 AmqpOutboundEndpoint 对象。对象 delegate 调用 handlerMessage 方法,handlerMessage 方法最终调用 handlerRequestMessage 方法。

handleRequestMessage 方法会生成 exchangeName 和 routingKey,然后根据是否需要返回值,来调用 convertAndSend 方法或 send 方法,如下所示。

```
// AmqpOutboundEndpoint.java
private void send(String exchangeName, String routingKey, org.springframework.messaging.
Message<?> requestMessage, CorrelationData correlationData) {
    if (this.amqpTemplate instanceof RabbitTemplate) {
        MessageConverter converter = ((RabbitTemplate)this.amqpTemplate).getMessageConverter();
        org.springframework.amqp.core.Message amqpMessage = MappingUtils.
mapMessage(requestMessage, converter, this.getHeaderMapper(), this.getDefaultDeliveryMode(),
this.isHeadersMappedLast());
        this.addDelayProperty(requestMessage, amqpMessage);
        ((RabbitTemplate)this.amqpTemplate).send(exchangeName, routingKey, amqpMessage,
correlationData);
    } else {
        this.amqpTemplate.convertAndSend(exchangeName, routingKey, requestMessage.
getPayload(), (message) -> {
            this.getHeaderMapper().fromHeadersToRequest(requestMessage.getHeaders(), message.
getMessageProperties());
            return message;
        });
    }
}
```

RabbitTemplate 是简化与 RabbitMQ 消息交互的工具类,它封装了 RabbitMQ 的客户端原生 API。其 send 方法最终会调用 doSend 方法。RabbitTemplate 的 doSend 方法则会调用 sendToRabbit 方法,进行真正的消息发送。sendToRabbit 方法调用了 RabbitMQ 中 Channel 对象的原生 API basicPublish 方法,将消息发送给 RabbitMQ 消息队列,如下所示。

```
protected void sendToRabbit(Channel channel, String exchange, String routingKey, boolean
 mandatory, org.springframework.amqp.core.Message message) throws IOException {
    AMQP.BasicProperties convertedMessageProperties = this.messagePropertiesConverter.
fromMessageProperties(message.getMessageProperties(), this.encoding);
    channel.basicPublish(exchange, routingKey, mandatory, convertedMessageProperties, message.
getBody());
}
```

9.5.3 @StreamListener 注解的处理

目前已经了解了输出通道初始化和消息发送的过程，下面就来看一看输入通道初始化和消息接收的过程。

@StreamListener 是一个可以修饰函数的注解，被它修饰的函数会用来接收输入通道的消息。下面就来探究一下 Spring Cloud Stream 是如何将消息队列 MessageInput.INPUT_MESSAGE 输入通道传来的消息分配到对应的函数的。

Spring Cloud Stream 定义了 StreamListenerAnnotationBeanPostProcessor 类来处理注解 @StreamListener，StreamListenerAnnotationBeanPostProcessor 实现了 BeanPostProcessor 接口，用来在 Bean 实例初始化之前和之后两个时间点对 Bean 实例进行处理。postProcessAfterInitialization 是在 Bean 实例初始化之后被调用的方法，它会遍历 Bean 实例中的所有函数，处理那些被 @StreamListener 注解修饰的函数，代码如下所示。

```java
// StreamListenerAnnotationBeanPostProcessor.java
public final Object postProcessAfterInitialization(Object bean, final String beanName) throws BeansException {
    Class<?> targetClass = AopUtils.isAopProxy(bean) ? AopUtils.getTargetClass(bean) : bean.getClass();
    Method[] uniqueDeclaredMethods = ReflectionUtils.getUniqueDeclaredMethods(targetClass);
    Method[] var5 = uniqueDeclaredMethods;
    int var6 = uniqueDeclaredMethods.length;

    for(int var7 = 0; var7 < var6; ++var7) {
        Method method = var5[var7];
        StreamListener streamListener = (StreamListener) AnnotatedElementUtils.findMergedAnnotation(method, StreamListener.class);
// 判断该 method 是否被@ StreamListener 修饰
        if (streamListener != null && !method.isBridge()) {
// 获取 streamListener 的基本信息
            this.streamListenerCallbacks.add(() -> {
                Assert.isTrue(method.getAnnotation(Input.class) == null, "A method annotated with @StreamListener may never be annotated with @Input. If it should listen to a specific input, use the value of @StreamListener instead");
                this.doPostProcess(streamListener, method, bean);
            });
        }
    }

    return bean;
}
```

registerHandlerMethodOnListenedChannel 方法会将 Method 和 @StreamListener 注解的相关信息组装成 StreamListenerHandlerMethodMapping 对象存储到 StreamListenerAnnotationBeanPostProcessor 的 mappedListenerMethods 对象中，供接下来统一处理时使用，代码如下所示。

```java
private void registerHandlerMethodOnListenedChannel(Method method, StreamListener streamListener, Object bean) {
    Assert.hasText(streamListener.value(), "The binding name can not be null");
    if (!StringUtils.hasText(streamListener.value())) {
```

```
        throw new BeanInitializationException("A bound component name must be specified");
    } else {
        String defaultOutputChannel = StreamListenerMethodUtils.getOutboundBindingTargetName(method);
        if (Void.TYPE.equals(method.getReturnType())) {
            Assert.isTrue(StringUtils.isEmpty(defaultOutputChannel), "An output channel can not be specified for a method that does not return a value");
        } else {
            Assert.isTrue(!StringUtils.isEmpty(defaultOutputChannel), "An output channel must be specified for a method that can return a value");
        }

        StreamListenerMethodUtils.validateStreamListenerMessageHandler(method);
            StreamListenerAnnotationBeanPostProcessor.this.mappedListenerMethods.add(streamListener.value(), StreamListenerAnnotationBeanPostProcessor.this.new StreamListenerHandlerMethodMapping(bean, method, streamListener.condition(), defaultOutputChannel, streamListener.copyHeaders()));
    }
}
```

StreamListenerAnnotationBeanPostProcessor 还实现了 SmartInitializingSingleton 接口，Spring 容器会在单例类型的实例创建结束时调用所有 SmartInitializingSingleton 接口的实现类的 afterSingletonsInstantiated 方法。StreamListenerAnnotationBeanPostProcessor 在 afterSingletonsInstantiated 方法中，对之前的 postProcessAfterInitialization 方法收集到的有关 @StreamListener 的信息进行处理。

afterSingletonsInstiantiated 方法会遍历 mappedListenerMethods 对象的所有 entry，为每一个 StreamListenerHandlerMethodMapping 创建一个 StreamListenerMessageHandler 实例，然后根据是否需要条件处理，生成 DispatchingStreamListenerMessageHandler 的 ConditionalStreamListenerHandler 实例。接着生成一个 DispatchingStreamListenerMessageHandler 实例，将之前生成的所有 ConditionalStreamListenerHandler 实例传给它，最后根据 mappedBindingEntry 的 key 值，也就是用 @StreamListener 的 value 值来获取 SubscribableChannel 实例，并调用其 subscribe 方法，将 DispatchingStreamListenerMessageHandler 实例注册给 SubscribableChannel。afterSingletonsInstantiated 方法的具体实现如下所示。

```
public final void afterSingletonsInstantiated() {
    this.injectAndPostProcessDependencies();
    EvaluationContext evaluationContext = IntegrationContextUtils.getEvaluationContext(this.applicationContext.getBeanFactory());
    Iterator var2 = this.mappedListenerMethods.entrySet().iterator();

    while(var2.hasNext()) {
        Map.Entry<String, List<StreamListenerAnnotationBeanPostProcessor.StreamListenerHandlerMethodMapping>> mappedBindingEntry = (Map.Entry)var2.next();
        ArrayList<DispatchingStreamListenerMessageHandler.ConditionalStreamListenerMessageHandlerWrapper> handlers = new ArrayList();
        Iterator var5 = ((List)mappedBindingEntry.getValue()).iterator();

        while(var5.hasNext()) {
            StreamListenerAnnotationBeanPostProcessor.StreamListenerHandlerMethodMapping mapping = (StreamListenerAnnotationBeanPostProcessor.StreamListenerHandlerMethodMapping)var5.next();
```

```java
// 使用函数 AOP，是为了将 Message 对象转化成函数输入参数的类型
        InvocableHandlerMethod invocableHandlerMethod = this.messageHandlerMethodFactory.createInvocableHandlerMethod(mapping.getTargetBean(), this.checkProxy(mapping.getMethod(), mapping.getTargetBean()));
        StreamListenerMessageHandler streamListenerMessageHandler = new StreamListenerMessageHandler(invocableHandlerMethod, this.resolveExpressionAsBoolean(mapping.getCopyHeaders(), "copyHeaders"), this.springIntegrationProperties.getMessageHandlerNotPropagatedHeaders());
        streamListenerMessageHandler.setApplicationContext(this.applicationContext);
        streamListenerMessageHandler.setBeanFactory(this.applicationContext.getBeanFactory());
// 处理 streamListener 的 condition 参数，通过 SpEL 解释器来解析 condition 的 value 值，然后新
建一个 ConditionalStreamListenerHandler
        if (StringUtils.hasText(mapping.getDefaultOutputChannel())) {
                streamListenerMessageHandler.setOutputChannelName(mapping.getDefaultOutputChannel());
        }

        streamListenerMessageHandler.afterPropertiesSet();
        if (StringUtils.hasText(mapping.getCondition())) {
            String conditionAsString = this.resolveExpressionAsString(mapping.getCondition(), "condition");
            Expression condition = SPEL_EXPRESSION_PARSER.parseExpression(conditionAsString);
            handlers.add(new DispatchingStreamListenerMessageHandler.ConditionalStreamListenerMessageHandlerWrapper(condition, streamListenerMessageHandler));
        } else {
            handlers.add(new DispatchingStreamListenerMessageHandler.ConditionalStreamListenerMessageHandlerWrapper((Expression)null, streamListenerMessageHandler));
        }
    }

    if (handlers.size() > 1) {
        var5 = handlers.iterator();

        while(var5.hasNext()) {
            DispatchingStreamListenerMessageHandler.ConditionalStreamListenerMessageHandlerWrapper handler = (DispatchingStreamListenerMessageHandler.ConditionalStreamListenerMessageHandlerWrapper)var5.next();
            Assert.isTrue(handler.isVoid(), "If multiple @StreamListener methods are listening to the same binding target, none of them may return a value");
        }
    }

    Object handler;
    if (handlers.size() <= 1 && ((DispatchingStreamListenerMessageHandler.ConditionalStreamListenerMessageHandlerWrapper)handlers.get(0)).getCondition() == null) {
        handler = ((DispatchingStreamListenerMessageHandler.ConditionalStreamListenerMessageHandlerWrapper)handlers.get(0)).getStreamListenerMessageHandler();
    } else {
        handler = new DispatchingStreamListenerMessageHandler(handlers, evaluationContext);
    }

    ((AbstractReplyProducingMessageHandler)handler).setApplicationContext(this.applicationContext);
    ((AbstractReplyProducingMessageHandler)handler).setChannelResolver(this.
```

```
            binderAwareChannelResolver);
        ((AbstractReplyProducingMessageHandler)handler).afterPropertiesSet();
            this.applicationContext.getBeanFactory().registerSingleton(handler.getClass().
getSimpleName() + handler.hashCode(), handler);
// 把之前生成的 DispatchingStreamListenerMessageHandler 实例注册到 SubscribableChannel 对象上
        ((SubscribableChannel)this.applicationContext.getBean((String)mappedBindingEntry.getKey(),
SubscribableChannel.class)).subscribe((MessageHandler)handler);
    }

    this.mappedListenerMethods.clear();
}
```

applicationContext 的 getBean 方法在获取 SubscribableChannel 实例时，就会用到 BindingBeansRegistrar 中 registerBindingTargetBeanDefinitions 方法注册的 BeanDefinition，最后由 BindableProxyFactory 来生成 SubscribableChannel 实例。

当 SubscribableChanel 接收到消息时，会调用 DispatchingStreamListenerMessageHandler 的 handleRequestMessage 方法。该方法会调用 findMatchingHandlers 方法来获取所有适合处理这个 Message 的 ConditionalStreamListenerHandler 实例，然后调用每个 ConditionalStreamListenerHandler 的 handleMessage 方法。handleRequestMessage 方法的实现如下所示。

```
protected Object handleRequestMessage(org.springframework.messaging.Message<?> requestMessage) {
    List<DispatchingStreamListenerMessageHandler.ConditionalStreamListenerMessageHandler
Wrapper> matchingHandlers = this.evaluateExpressions ? this.findMatchingHandlers(requestMessa
ge) : this.handlerMethods;
    if (matchingHandlers.size() == 0) {
        if (this.logger.isWarnEnabled()) {
            this.logger.warn("Can not find a @StreamListener matching for message with id: " +
requestMessage.getHeaders().getId());
        }

        return null;
    } else if (matchingHandlers.size() <= 1) {
        DispatchingStreamListenerMessageHandler.ConditionalStreamListenerMessageHandlerWrapp
er singleMatchingHandler = (DispatchingStreamListenerMessageHandler.ConditionalStreamListene
rMessageHandlerWrapper)matchingHandlers.get(0);
            singleMatchingHandler.getStreamListenerMessageHandler().handleMessage
(requestMessage);
        return null;
    } else {
        Iterator var3 = matchingHandlers.iterator();

        while(var3.hasNext()) {
            DispatchingStreamListenerMessageHandler.ConditionalStreamListenerMessageHandlerWra
pper matchingMethod = (DispatchingStreamListenerMessageHandler.ConditionalStreamListenerMe
ssageHandlerWrapper)var3.next();
            matchingMethod.getStreamListenerMessageHandler().handleMessage(requestMessage);
        }

        return null;
    }
```

}

如下方代码所示，findMatchingHandlers 方法主要是根据 ConditionalStreamListenerHandler 的条件判断的 Expression 实例来判断 ConditionalStreamListenerHandler 是否适合处理当前这个 Message 的。如果 ConditionalStreamListenerHandler 的 getCondition 方法返回的 Expression 实例为 null，那么认为该 ConditionalStreamListenerHandler 适合处理该 Message，否则调用其 getValue 方法来判断是否适合处理该 Message。

```java
private List<DispatchingStreamListenerMessageHandler.ConditionalStreamListenerMessageHandlerWrapper> findMatchingHandlers(org.springframework.messaging.Message<?> message) {
    ArrayList<DispatchingStreamListenerMessageHandler.ConditionalStreamListenerMessageHandlerWrapper> matchingMethods = new ArrayList();
    Iterator var3 = this.handlerMethods.iterator();

    while(var3.hasNext()) {
        DispatchingStreamListenerMessageHandler.ConditionalStreamListenerMessageHandlerWrapper conditionalStreamListenerMessageHandlerWrapperMethod = (DispatchingStreamListenerMessageHandler.ConditionalStreamListenerMessageHandlerWrapper)var3.next();
        if (conditionalStreamListenerMessageHandlerWrapperMethod.getCondition() == null) {
            matchingMethods.add(conditionalStreamListenerMessageHandlerWrapperMethod);
        } else {
            boolean conditionMetOnMessage = (Boolean)conditionalStreamListenerMessageHandlerWrapperMethod.getCondition().getValue(this.evaluationContext, message, Boolean.class);
            if (conditionMetOnMessage) {
                matchingMethods.add(conditionalStreamListenerMessageHandlerWrapperMethod);
            }
        }
    }

    return matchingMethods;
}
```

ConditionalStreamListenerHandler 的 handleMessage 方法就是直接调用它的 StreamListenerMessageHandler 成员变量的 handleMessage 方法的。

StreamListenerMessageHandler 是 AbstractReplyProducingMessageHandler 的子类，其 handleMessage 方法最终会调用自己声明的 handleRequestMessage 方法。

handleRequestMessage 方法调用了 invocableHandlerMethod 的 invoke 方法来调用对应的由 @StreamListener 注解修饰的方法，如下所示。

```java
// StreamListenerMessageHandler.java
protected Object handleRequestMessage(org.springframework.messaging.Message<?> requestMessage) {
    try {
        return this.invocableHandlerMethod.invoke(requestMessage, new Object[0]);
    } catch (Exception var3) {
        if (var3 instanceof MessagingException) {
            throw (MessagingException)var3;
        } else {
```

```
            throw new MessagingException(requestMessage, "Exception thrown while invoking " +
this.invocableHandlerMethod.getShortLogMessage(), var3);
        }
    }
}
```

其中，invocableHandlerMethod 对象是在 StreamListenerAnnotationBeanPostProcessor 的 afterSingletonsInstantiated 方法中创建并赋值给 StreamListenerMessageHandler 的。invocableHandlerMethod 对象包含了由@ StreamListener 修饰的函数和对应的 Bean 实例。

图 9-25 展示的是消息接收的流程，从 SubscribableChannel 接收到 RabbitMQ 传递过来的消息后，会调用 DispatchingStreamListenerMessageHandler 进行消息分发，这个处理器会根据 ConditionalStreamListenerHandler 的条件判断结果来传递消息，最终消息经过 invocableHandlerMethod 传递给 TestController 的对应函数。

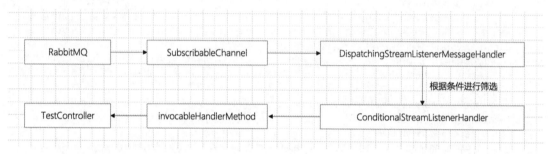

图9-25 消息接收流程

第 10 章 Spring Cloud Config 分布式配置中心

传统的单体应用常使用配置文件（比如 Spring Boot 的 application.yml 文件），来管理所有配置，但是在微服务架构中全部手动修改配置会很烦琐而且不易维护。这个问题可以使用 Spring Cloud Config 分布式配置中心来解决。

本章的主要内容如下。

1. 认识 Spring Cloud Config。
2. 实现配置中心。
3. 服务总线。
4. 源码解析。

10.1 认识 Spring Cloud Config

Spring Cloud Config 项目是一个分布式系统的配置管理解决方案。它包含了 Client 和 Server 两个部分，Server 提供配置文件的存储、以接口的形式将配置文件的内容提供出去；Client 通过接口获取数据、并依据此数据初始化自己的应用。

Spring Cloud Config 是 Spring Cloud 团队创建的一个全新项目，用来为分布式系统中的基础设施和微服务应用提供集中化的外部配置支持，它分为服务端与客户端两个部分。其中服务端也称为分布式配置中心（以下简称配置中心），它是一个独立的微服务应用，用来连接配置仓库并为客户端提供获取配置信息、加密/解密信息等访问接口；而客户端则是微服务架构中的各个微服务应用或基础设施，它们通过指定的配置中心来管理应用资源与业务相关的配置信息，并在启动的时候从配置中心获取和加载配置信息。

Spring Cloud Config 实现了对服务端和客户端中环境变量和属性配置的抽象映射，所以它除了适用于 Spring 构建的应用程序，还可以在任何其他语言构建的应用程序中使用。由于 Spring Cloud Config 实现的配置中心默认采用 Git 来存储配置信息，因此使用 Spring Cloud Config 构建的配置服务器，天然就支持对微服务应用配置信息的版本管理，并且可以通过 Git 客户端工具方便地管理和访问配置信息。当然它也提供了对其他存储方式的支持，比如 SVN 仓库、本地化文件系统等。

使用 Spring Cloud Config 配置中心后的架构如图 10-1 所示。

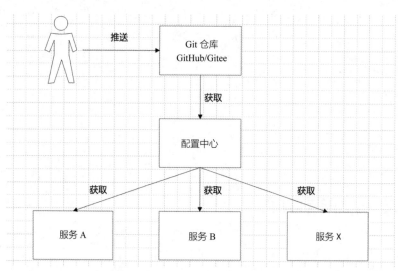

图10-1　使用Spring Cloud Config配置中心后的架构

需要注意的是，配置中心本质上也是一个微服务，同样需要注册到 Eureka 注册中心才能使用。

Spring Cloud Config 服务端的特性如下所示。

（1）使用 HTTP 为外部配置提供基于资源的 API（键值对或者等价的 YAML 内容）。

（2）属性值的加密和解密（对称加密和非对称加密）。

（3）通过使用 @EnableConfigServer 可以非常简单地嵌入 Spring Boot 应用。

Spring Cloud Config 客户端的特性如下所示。

（1）绑定 Spring Cloud Config 服务端，并使用远程的属性源初始化 Spring 环境。

（2）属性值的加密和解密（对称加密和非对称加密）。

10.1.1 配置中心概述

传统的单体应用，常使用配置文件（比如 Spring Boot 的 application.yml 文件）来管理所有配置，但是在微服务架构中全部手动修改配置会很繁琐而且不易维护。微服务的配置管理一般有以下需求。

（1）集中配置管理：一个微服务架构中可能有成百上千个微服务，所以集中配置管理是很重要的。

（2）不同环境不同配置：数据源配置在不同环境（开发、生成、测试环境）中是不同的。

（3）运行期间可动态调整：可根据各个微服务的负载情况，动态调整数据源连接池大小等。

（4）配置修改后可自动更新：配置信息发生变化，微服务可以自动更新配置。

综上所述对于微服务架构而言，一套统一的、通用的管理配置机制是不可缺少的重要组成部分。管理配置常见的做法就是通过配置服务器进行管理。

10.1.2 其他配置中心

除了 Spring Cloud Config，还有一些常见的配置中心，如下所示。

（1）Apollo。

Apollo（阿波罗）能够集中管理不同环境、不同集群的配置，配置修改后能够实时推送到应用端，并且具备规范的权限、流程治理等特性，适用于微服务配置管理场景。

（2）Disconf。

Disconf 是专注于各种分布式系统配置管理的通用组件和通用平台。提供统一的配置管理服务，有很多知名互联网公司正在使用。Disconf 在 2015 年度新增开源软件排名 TOP 100（中文开源技术交流社区提供）中排名第 16。

10.2 实现配置中心

以 Eureka 为注册中心，访问商品微服务，通过配置中心读取配置文件，端口号分别设置为 9001、9002 和 10000。

10.2.1 配置管理

Config Server 是一个可横向扩展、集中式的配置服务器，它用于集中管理应用程序在各个环境下的配置，默认使用 Git 存储配置文件，也可以使用 SVN 存储或者本地存储，这里使用 Git 作为学习的环境。

使用 GitHub 时，国内的用户经常遇到的问题是访问速度太慢，有时候还会出现无法连接的情况。如果使用者希望有一个更快速度的体验，可以使用国内的 Git 托管服务码云。和 GitHub 相比，码云也提供免费的 Git 仓库。此外，码云还集成了代码质量检测、项目演示等功能。对于团队协作开发，码云还提供了项目管理、代码托管和文档管理的服务。

■ 注册用户

打开码云的官方网站，找到注册入口并根据要求完成用户注册，如图 10-2 所示。

图10-2　注册用户

■ 创建仓库

注册后登录码云管理控制台，创建仓库，如图 10-3 所示。

图10-3　创建仓库

上传配置文件

修改子工程 product-service 的配置文件 application.yml 名称为 product-dev.yml，表示开发环境，内容如下所示。

```yaml
server: # 配置服务器
  port: 9001
spring:
  application:
    name: product-service
  datasource:
    driver-class-name: com.mysql.jdbc.Driver
    url: jdbc:mysql://mysql_server:3306/shop?characterEncoding=utf8
    username: root
    password: root
  jpa:
    database: MySQL
    show-sql: true
    open-in-view: true
eureka: # 配置 Eureka
  client:
    service-url:
      defaultZone: http://10.211.55.12:9000/eureka/
  instance:
    prefer-ip-address: true
    instance-id: ${spring.cloud.client.ip-address}:${server.port}
myname: myserver9001
```

修改子工程 product-service 的配置文件 application.yml 名称为 product-pro.yml，表示生产环境，内容如下所示。

```yaml
server: # 配置服务器
  port: 9002
spring:
  application:
    name: product-service
  datasource:
    driver-class-name: com.mysql.jdbc.Driver
    url: jdbc:mysql://mysql_server:3306/shop?characterEncoding=utf8
    username: root
    password: root
  jpa:
    database: MySQL
    show-sql: true
    open-in-view: true
eureka: # 配置 Eureka
  client:
    service-url:
      defaultZone: http://10.211.55.12:9000/eureka/
  instance:
    prefer-ip-address: true
    instance-id: ${spring.cloud.client.ip-address}:${server.port}
myname: myserver9002
```

开发者在项目发布之前，一般需要频繁地在开发环境、测试环境以及生产环境之间进行切换，这个时候大量的配置需要频繁更改（比如数据库、Redis、MongoDB、MQ 等配置）。频繁修改带来了巨大的工作量，Spring Boot 中约定的不同环境下配置文件命名规则（以区分开发环境、测试环境、生产环境等）如下所示。

```
{application}-{profile}.properties
或者
{application}-{profile}.yml
```

其中 application 为应用名称，profile 指的是开发环境。

将这两个文件上传到仓库中，如图 10-4 所示。

图10-4　上传文件

10.2.2　服务端

配置中心分为客户端和服务端，客户端通过服务端对外提供的访问路由获取配置文件信息。

■ 创建子工程

使用 IntelliJ IDEA 创建子工程——配置中心服务器（config-server），如图 10-5 所示。

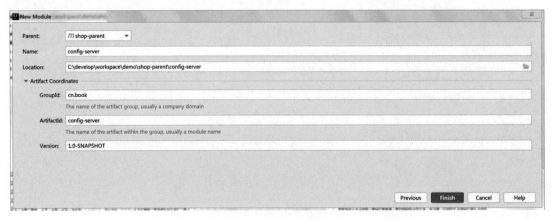

图10-5　创建配置中心服务器

目前的项目结构如图 10-6 所示。

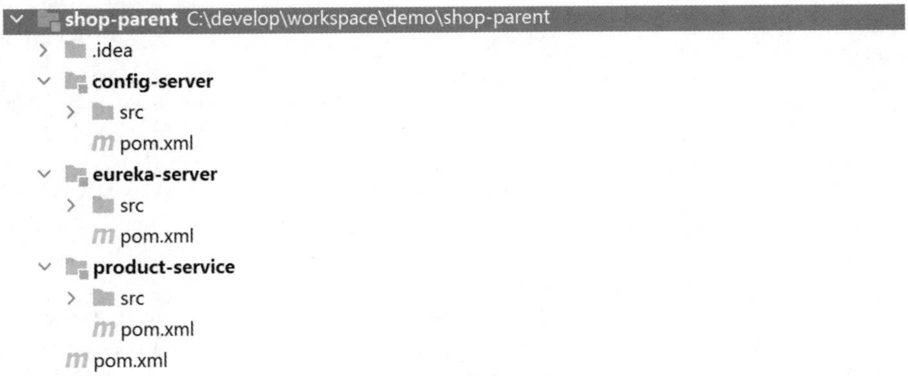

```
v  shop-parent  C:\develop\workspace\demo\shop-parent
   >  .idea
   v  config-server
      >  src
         pom.xml
   v  eureka-server
      >  src
         pom.xml
   v  product-service
      >  src
         pom.xml
      pom.xml
```

图10-6　项目结构

■ 引入依赖

在子工程 config-server 的 pom.xml 文件中引入配置中心相关依赖，如下所示。

```xml
<dependencies>
  <dependency>
    <groupId>org.springframework.cloud</groupId>
    <artifactId>spring-cloud-starter-netflix-eureka-client</artifactId>
  </dependency>
  <dependency>
    <groupId>org.springframework.cloud</groupId>
    <artifactId>spring-cloud-config-server</artifactId>
  </dependency>
</dependencies>
```

引入 Eureka 客户端依赖是因为 ConfigServer 也是一个微服务，需要注册到 Eureka。

■ 创建启动类

创建子工程 config-server 的启动类，如下所示。

```java
@SpringBootApplication
@EnableConfigServer
public class ConfigServerApplication {

    public static void main(String[] args) {
        SpringApplication.run(ConfigServerApplication.class,args);
    }
}
```

■ 创建配置文件

创建子工程 config-server 的配置文件 application.yml，如下所示。

```yaml
server:
  port: 10000
```

```yaml
spring:
  application:
    name: config-server
  cloud:
    config:
      server:
        git:
          uri: https://gitee.com/mybookconfig/myconfigcenter.git
eureka:
  client:
    service-url:
      defaultZone: http://10.211.55.12:9000/eureka/
  instance:
    prefer-ip-address: true
    instance-id: ${spring.cloud.client.ip-address}:${server.port}
```

通过 spring.cloud.config.server.git.uri 配置 Git 服务地址。

通过 spring.cloud.config.server.git.username 配置 Git 用户名。

通过 spring.cloud.config.server.git.password 配置 Git 密码。

■ 代码测试

按照顺序重启 EurekaServer 和 ConfigServer，访问服务注册中心管理后台，结果如图 10-7 所示。

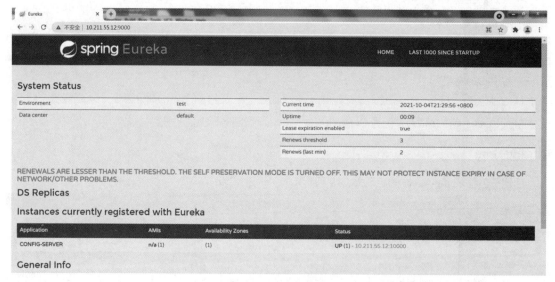

图10-7 服务注册中心管理后台

在图 10-7 中，发现 CONFIG-SERVER 已经作为服务注册到了服务注册中心管理后台中。接着访问 ConfigServer 的 product-dev.yml 和 product-pro.yml 配置文件，结果如图 10-8 和图 10-9 所示。

图10-8 访问product-dev.yml　　　　图10-9 访问product-pro.yml

在图 10-8 和图 10-9 中，访问路径中的 /product-pro.yml 和 /product-dev.yml 是 GitHub 中 sh 上传的文件的名字。

同时，可以看到 config-server 的控制台中输出了日志，如下所示。

```
........
2021-10-05 18:01:16.486  INFO 11384 --- [io-10000-exec-1] o.s.c.c.s.e.NativeEnvironmentRepository  : Adding property source: file:/C:/Users/admin/AppData/Local/Temp/config-repo-5583885994401456531/product-pro.yml
........
```

配置服务器在从 Git 中获取配置信息后，会存储一份在 ConfigServer 的文件系统中，实质上 ConfigServer 是通过 git clone 命令将配置信息的副本存储在本地，然后读取这些内容并返回给微服务应用进行加载的。

- **10.2.3　客户端**

这里以 product-service 作为配置中心客户端。

■ 引入依赖

在子工程 product-service 的 pom.xml 文件中引入配置中心相关依赖，如下所示。

```xml
<dependency>
  <groupId>org.springframework.cloud</groupId>
  <artifactId>spring-cloud-config-server</artifactId>
</dependency>
```

■ 配置文件

删除原来的配置文件 application.yml，使用加载级别更高的 bootstrap.yml 文件进行配置。启动应用时会检查此配置文件，在此配置文件中会指定配置中心的服务地址，自动地拉取所有应用配置并启用。创建配置文件 bootstrap.yml，内容如下所示。

```yaml
spring:
  cloud:
    config:
      name: product # 应用名称，需要对应 GitHub 中配置文件名称的前半部分
      profile: pro # 开发环境
      label: master #Git 中的分支
      uri: http://10.211.55.12:10000 #config-server 的请求地址
```

■ 实现高可用

这里 ConfigServer 的 URI 是硬编码到配置文件中的，不利于后期维护，另外，一旦 ConfigServer 宕机，势必会影响对应微服务的运行。可以启动多个 ConfigServer，服务名称保持一致（如果是在一台机器上需要设置不同的端口号），使用 Eureka 注册中心获取 ConfigServer 的配置。再次修改配置文件 bootstrap.yml，如下所示。

```yaml
spring:
  cloud:
    config:
      name: product # 应用名称，需要对应 GitHub 中配置文件名称的前半部分
      profile: pro # 开发环境
      label: master #Git 中的分支
      discovery: # 通过注册中心获取 config-server 配置
        enabled: true
        service-id: config-server
eureka: # 注册中心
  client:
    service-url:
      defaultZone: http://10.211.55.12:9000/eureka/
  instance:
    prefer-ip-address: true
    instance-id: ${spring.cloud.client.ip-address}:${server.port}
```

■ 修改控制器

修改子工程 product-service 的 ProductController，如下所示。

```java
@RestController
@RequestMapping("/product")
public class ProductController {
......
    @Value("${myname}")
    private String myname;

    @RequestMapping(value = "/{id}", method = RequestMethod.GET)
    public Product findById(@PathVariable Long id) {
        Product product = productService.findById(id);
        product.setProductName(" 访问的服务地址信息： "+ip + ":" + port+",myname="+myname);
        return product;
```

```
    }
  ……
}
```

代码测试

按照顺序启动 eureka-server、config-server 和 product-service，访问 ConfigServer Pro，如图 10-10 所示。

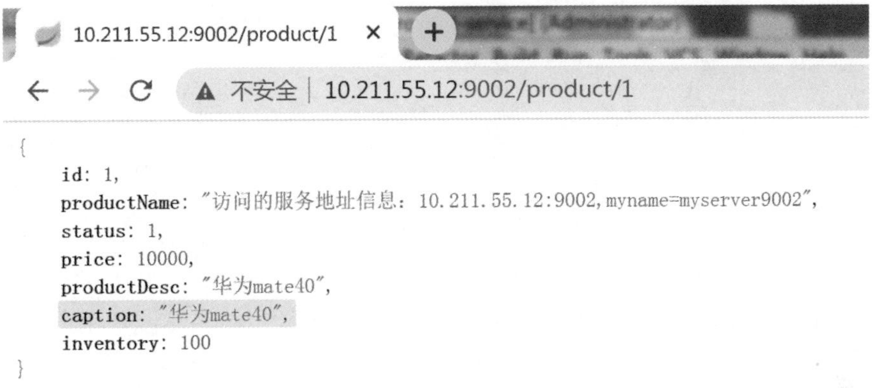

10.211.55.12:9002/product/1
不安全 | 10.211.55.12:9002/product/1
{
id: 1,
productName: "访问的服务地址信息：10.211.55.12:9002,myname=myserver9002",
status: 1,
price: 10000,
productDesc: "华为mate40",
caption: "华为mate40",
inventory: 100
}

图10-10 访问ConfigServer Pro

修改子工程 product-service 的配置文件 bootstrap.yml，将 pro 改为 dev，重新启动 product-service，访问 ConfigServer Dev，如图 10-11 所示。

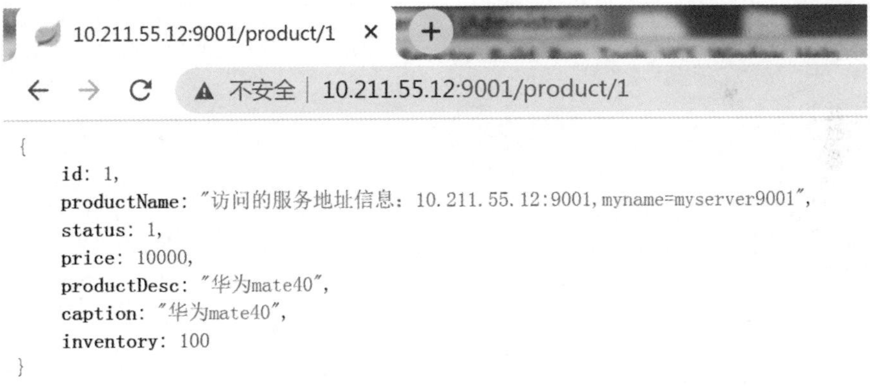

10.211.55.12:9001/product/1
不安全 | 10.211.55.12:9001/product/1
{
id: 1,
productName: "访问的服务地址信息：10.211.55.12:9001,myname=myserver9001",
status: 1,
price: 10000,
productDesc: "华为mate40",
caption: "华为mate40",
inventory: 100
}

图10-11 访问ConfigServer Dev

10.2.4 配置刷新

目前已经在客户端取到了配置中心的值，当修改 GitHub 上面的值时，服务端能实时获取最新的值。在图 10-12 所示的页面中单击【编辑】按钮进行编辑，将 myname 的值进行修改，如图 10-13 所示。

图10-12 编辑前

图10-13 编辑后

在图 10-13 所示的页面中，单击【提交】按钮后，通过 ConfigServer 访问配置文件，结果如图 10-14 所示。

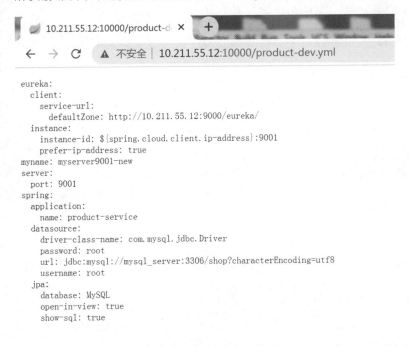

图10-14　通过ConfigServer访问配置文件

在图 10-14 中 myname 的值已经显示为修改后的值。接下来通过 ConfigClient 访问配置文件，结果如图 10-15 所示。

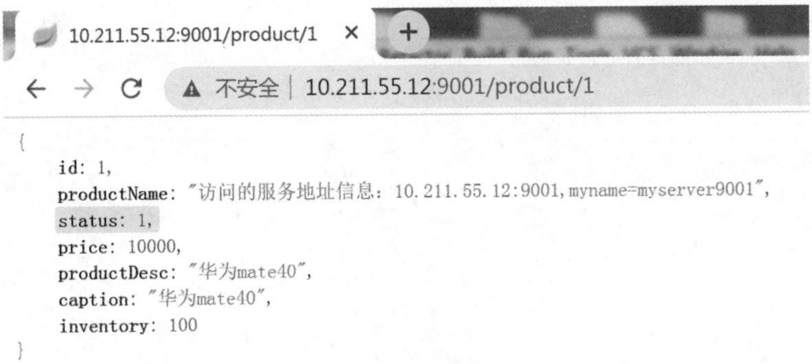

图10-15　通过ConfigClient访问配置文件

在图 10-15 中 ConfigClient 依然获取的是修改之前的值。

已经在客户端取到了配置中心的值，但当修改 GitHub 上面的值时，服务端能实时获取最新的值，但客户端是从缓存中读取数据的，无法实时获取最新的值。Spring Cloud 已经为使用者解决了这个问题，那就是客户端使用 POST 请求去刷新缓存，从而使获取最新数据。

■ 引入依赖

在子工程 product-service 的 pom.xml 文件中引入 Actuator 相关依赖，如下所示。

```xml
<dependency>
    <groupId>org.springframework.boot</groupId>
    <artifactId>spring-boot-starter-actuator</artifactId>
</dependency>
```

■ 修改控制器

修改子工程 product-service 的控制器 ProductController，添加刷新注解 @RefreshScope，如下所示。

```java
@RefreshScope
@RestController
@RequestMapping("/product")
public class ProductController {
......
}
```

■ 配置文件

修改子工程 product-service 的配置文件 bootstrap.yml，添加 Web 刷新路径，如下所示。

```yaml
spring:
  cloud:
    config:
      name: product
      profile: dev
      label: master
      discovery:
        enabled: true
        service-id: config-server
eureka:
  client:
    service-url:
      defaultZone: http://10.211.55.12:9000/eureka/
  instance:
    prefer-ip-address: true
    instance-id: ${spring.cloud.client.ip-address}:${server.port}

management:
  endpoints:
    web:
      exposure:
        include: refresh
```

■ 代码测试

再次修改服务端配置文件的内容，如下所示。

```
......
myname: myserver9001-new-new
......
```

重新启动 product-service，使用 Postman 向 http://10.211.55.12:9001/actuator/refresh 发送 POST 请求，如图 10-16 所示。

图10-16 使用Postman发送POST请求

POST 请求访问地址中的 /refresh 与配置文件 bootstrap.yml 中的 include:refresh 保持一致即可，当然也可以更换其他的单词。

此时查看客户端，可以看到输出的日志，如下所示。

```
……
2021-10-05 20:21:17.230  INFO 11384 --- [io-10000-exec-3] .c.s.e.MultipleJGitEnvironmentRepository : Fetched for remote master and found 1 updates
2021-10-05 20:21:17.339  INFO 11384 --- [io-10000-exec-3] o.s.c.c.s.e.NativeEnvironmentRepository  : Adding property source: file:/C:/Users/admin/AppData/Local/Temp/config-repo-5583885994401456531/product-dev.yml
……
```

通过输出的日志发现，已经刷新了本地缓存，接下来再次访问 product-server，结果如图 10-17 所示。

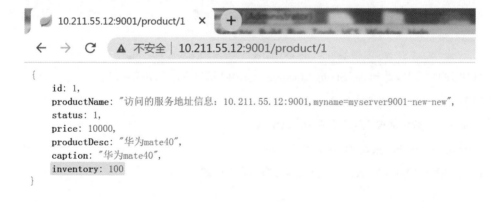

图10-17 再次访问结果

在图 10-17 中已经显示了修改之后的最新值，说明完成了配置刷新。

10.3 服务总线

在微服务架构的系统中，通常会使用轻量级的消息代理来构建一个共用的消息主题，让系统中所有微服务实例都连接起来。由于该主题中产生的消息会被所有实例监听和消费，因此称它为消息总线。

在消息总线上的各个实例都可以方便地广播一些需要让其他连接在该主题上的实例都知道的消息，例如配置信息的变更或者其他一些管理操作等。

由于消息总线在微服务架构系统中被广泛使用，因此它同配置中心一样，几乎是微服务架构中的必备组件。Spring Cloud 作为微服务架构综合性的解决方案，对此自然也有自己的实现，通过使用 Spring Cloud Bus 可以非常容易地搭建起消息总线，同时实现消息总线中的一些常用功能，比如配合 Spring Cloud Config 实现微服务应用配置信息的动态更新。通过 Spring Cloud Bus 动态更新配置的步骤如图 10-18 所示。

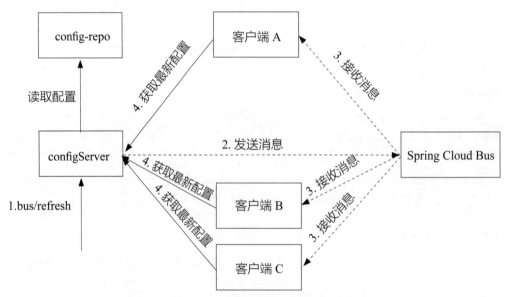

图10-18　通过Spring Cloud Bus动态更新配置的步骤

从图 10-18 中，可以看出利用 Spring Cloud Bus 动态更新配置的步骤如下所示。

（1）提交代码触发 POST 请求给 bus/refresh。

（2）服务端接收到请求并发送给 Spring Cloud Bus。

（3）Spring Cloud Bus 接收到消息并通知其他客户端。

（4）其他客户端接收到通知，向服务端请求获取最新配置。

10.3.1 消息代理

消息代理（Message Broker）是一种消息验证、传输、路由的架构模式。它在应用程序之间起到通信调度和最小化应用程序之间的依赖的作用，使得应用程序可以高效地解耦通信过程。消息代理是一个消息中间件，它的核心是一个消息的路由程序，用来实现接收和分发消息，并根据设定好的消息处理流来转发给正确的应用程序。它包括独立的通信和消息传递协议，能够实现组织内部和组织间的网络通信。设计消息代理就是为了能够从应用程序中传入消息，并执行一些特别的操作。

下面这些是在企业应用中经常需要使用消息代理的场景。

（1）将消息路由到一个或多个目的地。
（2）将消息转化为其他的表现形式。
（3）执行消息的聚集、消息的分解，并将结果发送到它们的目的地，然后重新组合响应
（4）返回给消息用户。
（5）调用 Web 服务来检索数据。
（6）响应事件或错误。
（7）使用发布订阅模式来提供内容或基于主题的消息路由。

目前已经有非常多的开源产品可以供大家使用，如下所示。

（1）ActiveMQ。
（2）Kafka。
（3）RabbitMQ。
（4）RocketMQ。

当前版本的 Spring Cloud Bus 仅支持两款消息中间件产品：RabbitMQ 和 Kafka。这里使用 RabbitMQ 与 Spring Cloud Bus 配合实现消息总线。

RabbitMQ 是实现了高级消息队列协议（AMQP）的开源消息代理软件，也称为面向消息的中间件。RabbitMQ 服务端是用因高性能、可伸缩而闻名的 Erlang 语言编写而成的，其集群和故障转移是构建在开放电信平台框架上的。

10.3.2　工程改造

以 10.2 节的代码为基础进行工程改造，工程包含 eureka-server、config-server 和 product-service。

服务端

修改服务端 config-server 的 pom.xml，引入相关依赖，如下所示。

```
<dependency>
    <groupId>org.springframework.cloud</groupId>
    <artifactId>spring-cloud-bus</artifactId>
</dependency>
<dependency>
    <groupId>org.springframework.cloud</groupId>
    <artifactId>spring-cloud-stream-binder-rabbit</artifactId>
</dependency>
```

修改服务端 config-server 的 application.yml，添加相关配置，如下所示。

```yaml
server:
  port: 10000
spring:
  application:
    name: config-server
  cloud:
    config:
      server:
        git:
          uri: https://gitee.com/mybookconfig/myconfigcenter.git
  rabbitmq:
    host: 127.0.0.1
    port: 5672
    username: guest
    password: guest
management: # 开启动态刷新的请求路径端点
  endpoints:
    web:
      exposure:
        include: bus/refresh
eureka:
  client:
    service-url:
      defaultZone: http://10.211.55.12:9000/eureka/
  instance:
    prefer-ip-address: true
    instance-id: ${spring.cloud.client.ip-address}:${server.port}
```

从上面的配置中，可以看到增加了以下两个参数。

（1）spring.rabbitmq：该参数指定消息中间件，需要提前开启 RabbitMQ。

（2）management：该参数指定开启动态刷新的请求路径端点。

■ 客户端

修改客户端 product-service 的 pom.xml，引入相关依赖，如下所示。

```xml
<dependency>
    <groupId>org.springframework.cloud</groupId>
    <artifactId>spring-cloud-bus</artifactId>
</dependency>
<dependency>
    <groupId>org.springframework.cloud</groupId>
    <artifactId>spring-cloud-stream-binder-rabbit</artifactId>
</dependency>
```

修改客户端 product-service 的 bootstrap.yml 中的相关配置，如下所示。

```yaml
spring:
  cloud:
    config:
      name: product
```

```
        profile: pro
        label: master
        discovery:
          enabled: true
          service-id: config-server
eureka:
  client:
    service-url:
      defaultZone: http://10.211.55.12:9000/eureka/
    instance:
      prefer-ip-address: true
      instance-id: ${spring.cloud.client.ip-address}:${server.port}
```

从上面的配置中，可以看到 spring.cloud.config.profile 的值设置为 pro。接着去修改 GitHub 中的配置文件 product-pro.yml，添加 RabbitMQ 的相关信息，如图 10-19 所示。

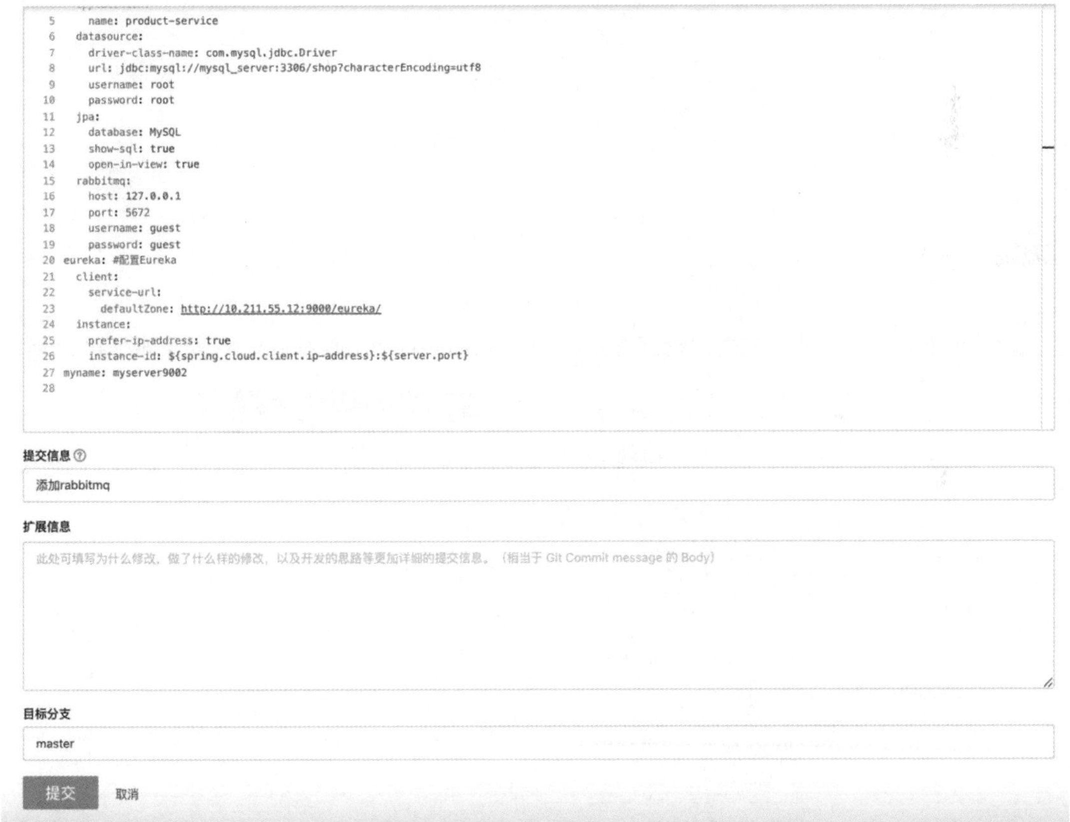

```
 5      name: product-service
 6    datasource:
 7      driver-class-name: com.mysql.jdbc.Driver
 8      url: jdbc:mysql://mysql_server:3306/shop?characterEncoding=utf8
 9      username: root
10      password: root
11    jpa:
12      database: MySQL
13      show-sql: true
14      open-in-view: true
15    rabbitmq:
16      host: 127.0.0.1
17      port: 5672
18      username: guest
19      password: guest
20  eureka: #配置Eureka
21    client:
22      service-url:
23        defaultZone: http://10.211.55.12:9000/eureka/
24    instance:
25      prefer-ip-address: true
26      instance-id: ${spring.cloud.client.ip-address}:${server.port}
27  myname: myserver9002
28
```

提交信息 ⑦
添加rabbitmq

扩展信息
此处可填写为什么修改，做了什么样的修改，以及开发的思路等更加详细的提交信息。（相当于 Git Commit message 的 Body）

目标分支
master

提交 取消

图10-19 修改product-pro.yml

重新启动服务端和客户端，修改 GitHub 中的 product-pro.yml 中 myname 的值为 myserver9002-bus，如图 10-20 所示。

```
11   jpa:
12     database: MySQL
13     show-sql: true
14     open-in-view: true
15   rabbitmq:
16     host: 127.0.0.1
17     port: 5672
18     username: guest
19     password: guest
20   eureka: #配置Eureka
21     client:
22       service-url:
23         defaultZone: http://10.211.55.12:9000/eureka/
24     instance:
25       prefer-ip-address: true
26       instance-id: ${spring.cloud.client.ip-address}:${server.port}
27   myname: myserver9002-bus
28
```

提交信息

update product-pro.yml.

扩展信息

此处可填写为什么修改，做了什么样的修改，以及开发的思路等更加详细的提交信息。（相当于 Git Commit message 的 Body）

目标分支

master

提交　取消

图10-20　修改myname的值

在图 10-20 所示的页面中，单击【提交】按钮后，通过 ConfigServer 访问配置文件，结果如图 10-21 所示。

图10-21　通过ConfigServer访问配置文件

在图 10-21 中 myname 的值已经显示为修改后的最新值。接下来通过 ConfigClient 访问配置文件，结果如图 10-22 所示。

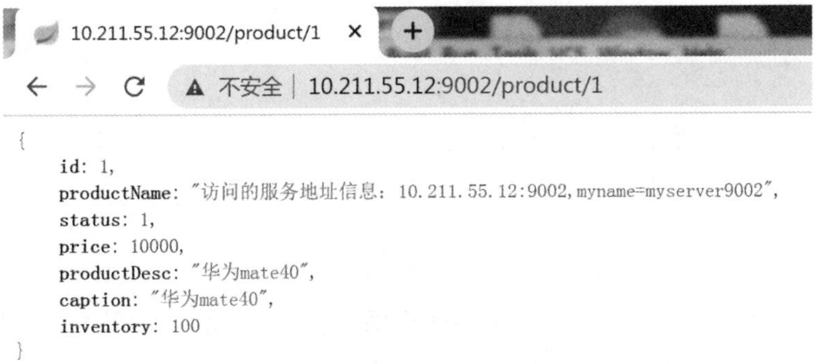

图10-22　通过ConfigClient访问配置文件

在图 10-22 中 ConfigClient 依然获取的是修改之前的值。接下来使用 Postman 向 http://10.211.55.12:10000/actuator/bus-refresh 发送 POST 请求，如图 10-23 所示。

图10-23　使用Postman发送POST请求

此时查看客户端，可以看到输出的日志，如下所示。

```
……
2021-10-08 19:24:39.146  INFO 6284 --- [io-10000-exec-2] .c.s.e.MultipleJGitEnvironmentRepository
: Fetched for remote master and found 1 updates
2021-10-08 19:24:39.239  INFO 6284 --- [io-10000-exec-2] o.s.c.c.s.e.NativeEnvironmen
tRepository   : Adding property source: file:/C:/Users/admin/AppData/Local/Temp/config-repo-7786712969161969798/product-pro.yml
……
```

通过输出的日志发现，已经刷新了本地缓存，再次访问配置文件，结果如图 10-24 所示。

图10-24 再次访问配置文件

在图 10-24 中已经显示了修改之后的最新值,说明完成了配置刷新。如果有多个客户端,也只需要执行一次 POST 请求,就可以完成所有客户端配置文件的刷新。

10.4 源码解析

配置中心包括如下元素。

(1)配置服务器:为配置客户端提供其对应的配置信息,配置信息的来源为配置仓库,启动时即拉取配置仓库的信息,并将其缓存到本地仓库中。

(2)配置客户端:除了配置服务器之外的应用服务,还在启动时从配置服务器拉取其对应的配置信息。

(3)配置仓库:为配置服务器提供配置源信息,配置仓库的实现可以支持多种方式。

配置中心的应用架构如图 10-25 所示。

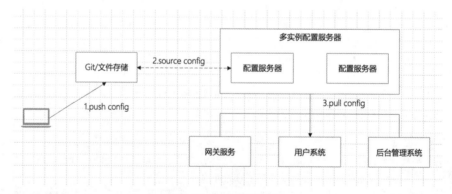

图10-25 配置中心的应用架构

在部署环境之前,需要将相应的配置信息推送到配置仓库,如图 10-26 所示。配置仓库支持多样的源,如 Git、SVN、JDBC 数据库和本地文件系统等。先启动配置服务器,启动之后,将配置信息拉取并同步至本地仓库。然后通过配置服务器对外提供的 REST 接口,其他所有的配置客户端在启动时根据 spring.cloud.config 配置的 {application}/ {profile}/{label} 信息去配置服务器拉取相应的配置。最后,其他客户端应用启动,从配置服务器拉取配置信息。配置中心支持动态刷新配置信息,即不需要重启应用,结合消息总线提供的刷新 API，通过 Webhook 调用该端点 API，达到动态刷新的效果。

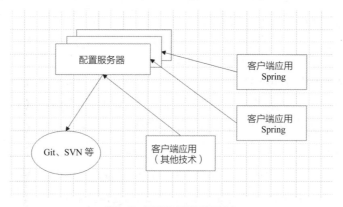

图10-26　将配置信息推送到配置仓库

总的来说，Spring Cloud Config 具有如下特性。

（1）提供配置服务器和配置客户端支持。

（2）集中管理分布式环境下的应用配置。

（3）基于 Spring 环境，可以无缝地与 Spring 应用集成。

（4）可用于任何语言开发的程序，为其提供与管理配置信息。

（5）默认实现基于 Git 仓库，可以进行版本管理。

下面分别对配置服务器和配置客户端的主要功能进行源码解析。

10.4.1　配置服务器

配置服务器的主要功能有连接配置仓库、拉取远端的配置并缓存到本地、对外提供配置信息的 REST 接口。下面围绕配置服务器的这几个主要功能进行介绍。

ConfigServer 配置类

在讲解配置服务器前，首先介绍 ConfigServer 配置类。注解 @EnableConfigServer 可以开启应用服务对配置中心的支持。

当启用了 Config Server 之后，配置服务器在启动时就需要对 ConfigServer 进行自动配置，在 ConfigServerAutoConfiguration 中引入了多个配置类，主要包括以下配置类。

（1）EnvironmentRepositoryConfiguration：环境变量存储相关的配置。

（2）CompositeConfiguration：组合方式的环境仓储配置。

（3）ResourceRepositoryConfiguration：资源仓储相关的配置。

（4）ConfigServerEncryptionConfiguration：加密端点的配置。

（5）ConfigServerMvcConfiguration：对外暴露的 MVC 端点控制器的配置。

（6）TransportConfiguration：Clone 或 Fetch 传输命令的回调配置。

@Import 注解导入了这些配置类，这些类也是 Config Server 的主要功能实现。下面重点介绍核心的配置类，通过配置类来引入 Config Server 核心功能。

■ 获取指定服务的环境配置

1. EnvironmentRepositoryConfiguration

EnvironmentRepositoryConfiguration 支持如下的配置仓库源：JDBC、Vault（HashiCorp 开发的一款私密信息管理工具）、SVN、本地文件系统、Git。EnvironmentRepositoryConfiguration 包括如下配置类。

（1）CompositeRepositoryConfiguration：组合的配置仓储配置类，对应的 Profile 为 Composite。

（2）JdbcRepositoryConfiguration：基于 JDBC 数据库存储的配置类，对应的 Profile 为 JDBC。

（3）VaultRepositoryConfiguration：Vault 方式的配置类，对应的 Profile 为 Vault。

（4）SvnRepositoryConfiguration：SVN 方式的配置类，对应的 Profile 为 Subversion。

（5）NativeRepositoryConfiguration：本地文件方式的配置类，对应的 Profile 为 Native。

（6）GitRepositoryConfiguration：Git 方式的配置类，对应的 Profile 为 Git。

（7）DefaultRepositoryConfiguration：默认方式的配置类，即不指定 Profile 时，与 Profile 为 Git 的方式相同。

环境仓库支持多种方式，每种方式的实现原理都是相同的。这里以其中最常用的默认方式为例进行讲解，即采用 Git 方式配置环境仓库。

既然环境仓库支持多种方式，那么怎么指定配置服务器启动时使用哪种方式呢？配置为 spring.cloud.config.server.git，但这里的配置并没看出指定了哪种配置仓库的方式，所以看一下该配置类的具体实现，如下所示。

```
@Configuration
@ConditionalOnMissingBean(
   value = {EnvironmentRepository.class},
   search = SearchStrategy.CURRENT
)
// 没有指定默认的环境仓库时，将会默认初始化
class DefaultRepositoryConfiguration {
......
   @Bean
    public MultipleJGitEnvironmentRepository defaultEnvironmentRepository(MultipleJGitEnvironmentRepositoryFactory gitEnvironmentRepositoryFactory, MultipleJGitEnvironmentProperties environmentProperties) throws Exception {
// 使用的是 Git 方式
      return gitEnvironmentRepositoryFactory.build(environmentProperties);
   }
}
```

EnvironmentRepositoryConfiguration 中声明了多个其他的配置类，上述代码只展示了默认方式实现代码。可以看出，每种配置仓库的实现都对应声明的配置类，用 @Profile 注解来激活相应的配置类，并在配置服务器的 application.properties 或 application.yml 中指定 spring.profiles.active 为 jdbc。当没有设置的时候，使用

的是默认的 Git 方式，所以看到 DefaultRepositoryConfiguration 配置的注解有 @ConditionalOnMissingBean (EnvironmentRepository.class)，当上下文中不存在 EnvironmentRepository 对象时才会实例化该默认的环境仓库配置类。

2. EnvironmentRepository 接口和 SearchPathLocator 接口

各类环境仓库都实现了顶级接口 EnvironmentRepository 和 SearchPathLocator。前者的定义如下所示。

```
public interface EnvironmentRepository {
    Environment findOne(String application, String profile, String label);
}
```

该接口定义了一个获取指定应用服务环境信息的方法 findOne，返回的是 Environment 对象。

findOne 需要传入的参数也很熟悉，有 application、profile 和 label，对应于客户端应用的信息。该方法根据传入的客户端应用的信息，获取对应的配置信息。

除了传入的客户端应用的信息，还有对应于 Git 提交的 CommitId，PropertySources 对应环境变量的源和具体的值。

另一个接口是 SearchPathLocator，根据传入的客户端应用信息，获取对应的配置环境文件的位置信息。该接口的定义如下所示。

```
public interface SearchPathLocator {
    SearchPathLocator.Locations getLocations(String application, String profile, String label);

    public static class Locations {
        private final String application;
        private final String profile;
        private final String label;
        private final String[] locations;
        private final String version;

        public Locations(String application, String profile, String label, String version, String[] locations) {
            this.application = application;
            this.profile = profile;
            this.label = label;
            this.locations = locations;
            this.version = version;
        }
        ......
}
```

Locations 对应的是本地仓库的位置，由于是根据 Profile 区分，因此这里返回的两个位置为根目录和 Dev 目录。SearchPathLocator 接口用于定位资源搜索路径，比如配置信息存储于文件系统或者类路径中，获取配置信息时需要准确定位到这些配置的位置。内部类 Locations 定义了应用服务配置存储信息。和 findOne 方法的参数一样，getLocations 方法也需要应用服务的相关信息。

3. JGit 方式

图 10-27 所示为接口 EnvironmentRepository 和 SearchPathLocator 的类图。

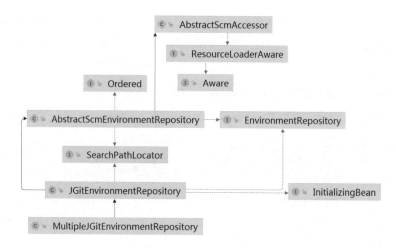

图10-27 接口EnvironmentRepository和SearchPathLocator的类图

MultipleJGitEnvironmentRepository 继承自 JGitEnvironmentRepository，JGit 也是一种配置环境仓储的方式。

从图 10-27 可看出，JGitEnvironmentRepository 继承自抽象类 AbstractScmEnvironmentRepository，而该抽象类又继承自 AbstractScmAccessor。AbstractScmAccessor 是 SCM（软件配置管理，Source Control Management）实现的父类，定义了基础的属性以获取 SCM 的资源。从基类 AbstractScmAccessor 看起，在基类中定义了 SCM 的配置属性和基本方法，包括远端仓库的地址、用户名、密码等属性，这些在使用 Git 仓库时都需要配置。还有一些属性，如 basedir（即本地复制之后的工作仓库地址）和 passphrase（即 SSH 的私钥）等，都是配置仓库时的可选项。

在抽象类 AbstractScmEnvironmentRepository 中，实现了在 EnvironmentRepository 接口中获取配置信息的方法，代码如下所示。

```java
// AbstractScmEnvironmentRepository.java
public synchronized Environment findOne(String application, String profile, String label) {
    NativeEnvironmentRepository delegate = new NativeEnvironmentRepository(this.getEnvironment(), new NativeEnvironmentProperties());
    SearchPathLocator.Locations locations = this.getLocations(application, profile, label);
    delegate.setSearchLocations(locations.getLocations());
    Environment result = delegate.findOne(application, profile, "");
    result.setVersion(locations.getVersion());
    result.setLabel(label);
    return this.cleaner.clean(result, this.getWorkingDirectory().toURI().toString(), this.getUri());
}
```

在获取应用服务的配置信息时，首先新建了一个本地的环境仓库，作为代理的环境仓库。然后获取本地环境仓库中指定应用服务的位置（一般是一个 Tmp 目录，也可以自行指定）。在获取到本地的搜索路径之后，根据该路径搜索应用服务的配置信息。最后将得到的结果进行处理，设置应用服务的 Profile 和标签等。由于 findOne 方法返回的 Environment 对象包含 WorkingDir 和 JSON 键的相关信息，因此这里还调用了 EnvironmentCleaner#clean 方法，对结果进行处理。该抽象类基于 SCM 对配置信息进行管理的子类为

JGitEnvironmentRepository。

下面具体介绍 JGit 中实现的方法，如下所示。

```java
// JGitEnvironmentRepository.java
public String refresh(String label) {
// 调用 Git 操作的方法，准备好工作目录，返回最新的 HEAD 版本号
......
}
......
public synchronized SearchPathLocator.Locations getLocations(String application, String profile, String label) {
    if (label == null) {
        label = this.defaultLabel;
    }

    String version = this.refresh(label);
    return new SearchPathLocator.Locations(application, profile, label, version, this.getSearchLocations(this.getWorkingDirectory(), application, profile, label));
}
```

JGitEnvironmentRepository 继承抽象类 AbstractScmEnvironmentRepository，在 SCM 的基础上加入了 Git 操作相关的方法。获取具体的配置文件地址时，需要传入应用名称、Profile 和标签信息，根据最新的版本号返回 Locations 定位到的资源搜索路径。refresh 方法用于刷新本地仓库的配置状态，保证每次都能拉取到最新的配置信息。

4. JGit 方式

JGit 方式作为 AbstractScmEnvironmentRepository 子类，并没有覆写 findOne，其获取配置信息的逻辑是：获取指定应用在本地仓库中的路径，根据获取的本地位置作为搜索路径，最后获取到本地仓库中的配置信息。

其中，第一步的逻辑极为重要，因为这关乎获取的配置信息是否是最新的。在获取本地仓库路径之前，需要检查 Git 仓库的状态，代码如下所示。

```java
// JGitEnvironmentRepository.java
public String refresh(String label) {
    Git git = null;

    String var20;
    try {
        git = this.createGitClient();
        if (this.shouldPull(git)) {
            FetchResult fetchStatus = this.fetch(git, label);
            if (this.deleteUntrackedBranches && fetchStatus != null) {
                this.deleteUntrackedLocalBranches(fetchStatus.getTrackingRefUpdates(), git);
            }
            this.checkout(git, label);
            this.tryMerge(git, label);
        } else {
            this.checkout(git, label);
            this.tryMerge(git, label);
        }
// 返回 HEAD 版本号
```

```
            var20 = git.getRepository().findRef("HEAD").getObjectId().getName();
        } catch (RefNotFoundException var15) {
            throw new NoSuchLabelException("No such label: " + label, var15);
        } catch (NoRemoteRepositoryException var16) {
            throw new NoSuchRepositoryException("No such repository: " + this.getUri(), var16);
        } catch (GitAPIException var17) {
            throw new NoSuchRepositoryException("Can not clone or checkout repository: " + this.getUri(), var17);
        } catch (Exception var18) {
            throw new IllegalStateException("Can not load environment", var18);
        } finally {
            try {
                if (git != null) {
                    git.close();
                }
            } catch (Exception var14) {
                this.logger.warn("Could not close git repository", var14);
            }

        }

        return var20;
    }
    ......
    protected boolean shouldPull(Git git) throws GitAPIException {
// 判断是否需要拉取
        if (this.refreshRate > 0 && System.currentTimeMillis() - this.lastRefresh < (long)(this.refreshRate * 1000)) {
            return false;
        } else {
// 获取远端 Git 仓库的状态
            Status gitStatus = git.status().call();
            boolean isWorkingTreeClean = gitStatus.isClean();
            String originUrl = git.getRepository().getConfig().getString("remote", "origin", "url");
            boolean shouldPull;
            if (this.forcePull && !isWorkingTreeClean) {
                shouldPull = true;
                this.logDirty(gitStatus);
            } else {
                shouldPull = isWorkingTreeClean && originUrl != null;
            }

            if (!isWorkingTreeClean && !this.forcePull) {
                this.logger.info("Can not pull from remote " + originUrl + ", the working tree is not clean.");
            }

            return shouldPull;
        }
    }
```

通过检查远端仓库的 Git 状态，进而判断本地仓库是否需要刷新。refresh 依赖于 shouldPull 的状态，当有新的提交或者配置了强制拉取时，Git 客户端会 Fetch 所有的更新，并 Merge 到所在分支或 Tag，更新本地仓库。refresh 最终返回的是最新一次提交的 HEAD 版本号。

5. 多个 JGit 仓库

ConfigServer 中还可以配置多个 Git 仓库，服务启动时会自动读取相关配置属性，MultipleJGitEnvironmentRepository 是 JGit 实现的子类。MultipleJGitEnvironmentRepository 用于处理一个或多个 Git 仓库的环境仓储，且 MultipleJGitEnvironmentRepository 会遍历所有的仓库。其具体实现和 JGit 方式差异并不是很大，这里不再展开描述。

■ 获取指定服务的资源文件

ResourceRepositoryConfiguration 在 SearchPathLocator 对象存在时，将 ResourceRepository 加入 Spring 的上下文，如下所示。

```java
// ResourceRepositoryConfiguration.java
@Bean
@ConditionalOnBean({SearchPathLocator.class})
public ResourceRepository resourceRepository(SearchPathLocator service) {
    return new GenericResourceRepository(service);
}
```

前面介绍了 EnvironmentRepository，相比而言，ResourceRepository 用来定位一个应用的资源，返回的是某一个具体的资源文件，将其内容转换成文本格式。而 EnvironmentRepository 返回的信息更加全面，信息采用键值对的格式，包括应用的基本信息和指定应用的配置源（可能来自多个源或者共享配置文件），这些键值对可以替换资源文件中的占位符。ResourceRepository 接口的定义如下所示。

```java
public interface ResourceRepository {
    Resource findOne(String name, String profile, String label, String path);
}
```

返回的 Resource 是一个资源描述符的接口，用于抽象底层资源的实际类型，如文件或类路径资源。

```yaml
spring:
profiles: dev
cloud:
version: Camden SR4
```

上面是示例项目中客户端服务获取其对应的资源文件的结果，Resource 流转化成 String 文本。ResourceRepository 接口的实现类为 GenericResourceRepository。GenericResourceRepository 覆写了 findOne 方法，返回配置信息。findOne 方法的实现如下所示。

```java
// GenericResourceRepository.java
public synchronized Resource findOne(String application, String profile, String label, String path) {
    String[] locations = this.service.getLocations(application, profile, label).getLocations();

    try {
        int i = locations.length;

        while(i-- > 0) {
            String location = locations[i];
```

```
            Iterator var8 = this.getProfilePaths(profile, path).iterator();

            while(var8.hasNext()) {
                String local = (String)var8.next();
                Resource file = this.resourceLoader.getResource(location).createRelative(local);
                if (file.exists() && file.isReadable()) {
                    return file;
                }
            }
        }
    } catch (IOException var11) {
        throw new NoSuchResourceException("Error : " + path + ". (" + var11.getMessage() + ")");
    }

    throw new NoSuchResourceException("Not found: " + path);
}
```

从上面的实现来看，该方法主要是通过构造方法设置 SearchPathLocator 对象，传入应用名称等参数调用 getLocations 方法得到配置源的具体路径的，SearchPathLocator 的实现类会保证配置仓库是最新的。因为 Profile 可以有默认值 Default，在创建资源文件之前，先调用 getProfilePaths 方法根据 Profile 对 path 进行处理，然后由 ResourceLoader#getResource 方法创建绝对路径的配置源。

■ ConfigServer 提供的端点

ConfigServerMvcConfiguration 对外提供的 API 端点包括 3 类：Environment、Resource 以及加密 / 解密的端点。下面介绍前两类常用的 API 端点。ConfigServerMvcConfiguration 的定义如下所示。

```java
// ConfigServerMvcConfiguration.java
@Bean
public EnvironmentController environmentController(EnvironmentRepository envRepository,
ConfigServerProperties server) {
    EnvironmentController controller = new EnvironmentController(this.encrypted(envRepository,
server), this.objectMapper);
    controller.setStripDocumentFromYaml(server.isStripDocumentFromYaml());
    controller.setAcceptEmpty(server.isAcceptEmpty());
    return controller;
}
......
private EnvironmentRepository encrypted(EnvironmentRepository envRepository,
ConfigServerProperties server) {
    EnvironmentEncryptorEnvironmentRepository encrypted = new EnvironmentEncryptorEnvironmentRepository(envRepository, this.environmentEncryptor);
    encrypted.setOverrides(server.getOverrides());
    return encrypted;
}
```

如上所示，配置文件 ConfigServerMvcConfiguration 将 EnvironmentController 加入了 Spring 的上下文，并自动注入 EnvironmentRepository 和 ConfigServerProperties 对象。在将配置信息返回给客户端服务之前，远端加密的属性值（以 {cipher} 开头的字符串）将会被解密。这里的 encrypted 方法，就是将给定的 EnvironmentRepository 再次封装，返回一个代理类，以解密属性值。Overrides 属性即在配置服务器设置的属性，用于强制覆写客户端对应的环境变量，这里一起封装到代理类中。StripDocumentFromYaml 属性用来标识那些不是 Map 类型的 YAML 文件，应该去掉 Spring 增加的文件的前缀，该属性的默认值为 true。

1. 获取 Environment 的 API 端点

Environment 控制器提供了如下的 API 端点。

（1）/{application}/ [(profile)] [/ (label)]。

（2）/{application}-{profile}. yml。

（3）/{label}/{application}-{profile} .yml。

（4）/{application}-{profile}. properties。

（5）/{label}/{application}-{profile}.properties。

如上的格式都可以获取指定应用的配置信息，如 URI 为 /config-client/dev 和 /config-client-dev.yml 等。配置仓库采用 Git 方式获取指定应用的 Environment 的过程如图 10-28 所示。

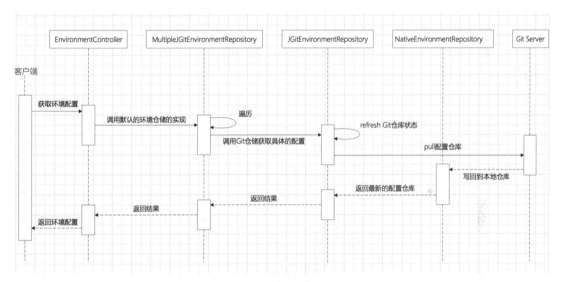

图10-28　获取指定应用的Environment过程

从图 10-28 中可以清楚地知道客户端每次拉取的配置都是本地仓库复制的那一份，通过 NativeEnvironmentRepository 代理获取指定应用的配置。ConfigServer 每次都会检查指定 Git 仓库的状态，当远端仓库有更新时，则会 Fetch 到本地进行更新。下面所示为 EnvironmentController 的实现。

```
// EnvironmentController.java
// 根据参数 name 和 profiles 返回 Environment 对象
@RequestMapping({"/{name}/{profiles:.*[^-].*}"})
public Environment defaultLabel(@PathVariable String name, @PathVariable String profiles) {
    return this.labelled(name, profiles, (String)null);
}
@RequestMapping({"/{label}/{name}-{profiles}.properties"})
public ResponseEntity<String> labelledProperties(@PathVariable String name, @PathVariable String profiles, @PathVariable String label, @RequestParam(defaultValue = "true") boolean resolvePlaceholders) throws IOException {
    this.validateProfiles(profiles);
```

```
    Environment environment = this.labelled(name, profiles, label);
    Map<String, Object> properties = this.convertToProperties(environment);
    String propertiesString = this.getPropertiesString(properties);
    if (resolvePlaceholders) {
        propertiesString = EnvironmentPropertySource.resolvePlaceholders(EnvironmentPropertySource.prepareEnvironment(environment), propertiesString);
    }

    return this.getSuccess(propertiesString);
}
```

控制器调用相应环境仓库实现的 FindOne 方法，既可以返回完整的 Environment 对象，包括 Name、Profiles 和 PropertySources 等信息，也可以直接返回配置仓库中的源配置数据，还可以直接返回处理之后的 JSON 对象。在 /{label}/{name}-{profiles}.properties 这个接口对应的实现中，首先调用 labelled 方法返回 Environment 对象，然后将该对象转换成 Map，由 Map 转换成 String，最后替换掉系统的环境变量的占位符，返回文本对象。

2. 获取 Resource 的 API 端点

Resource 提供的 API 端点实现其实有些类似 Environment API 端点的实现，在此仅简单介绍。下面所示为获取 Resource 的 API 端点。

（1）/{name}/{profile}/ {label}/**。

（2）/{name}/{profile}/**。

获取资源文件的过程和获取环境变量的过程类似，不同的是，当返回的是一个指定的配置文件时，Resource 控制器默认会将文件中的占位符替换，最后将替换后的资源文件返回给客户端。

Resource 控制器和 EnvironmentController 的初始化过程类似，Resource 控制器依赖的获取资源文件的 EnvironmentRepository 也是 Encryptor 代理类，会将解密后的属性值返回给客户端。获取具体应用的资源文件的方法，需要应用名称、Profile、标签和文件名，根据这些参数便可以得到相应的资源文件，将 Resource 流转换成字符串返回给客户端。默认会将资源文件中的系统环境变量占位符进行替换。

10.4.2 配置客户端

Spring Boot 的应用能够使开发者立刻体验 Spring Cloud Config 服务端带来的优势，还能够收获与环境变化事件有关的特性。在依赖中添加 spring-cloud-config-client 依赖，即可对 Spring Cloud Config 进行初始化配置。客户端应用在启动时有以下两种配置方式。

（1）通过 HTTP URI 指定 Config Server。

（2）通过服务发现指定 Config Server。

与之相关的配置类为在 spring.factories 文件中定义的启动上下文，如下所示。

```
# Bootstrap components
org.springframework.cloud.bootstrap.BootstrapConfiguration=\
org.springframework.cloud.config.client.ConfigServiceBootstrapConfiguration,\
org.springframework.cloud.config.client.DiscoveryClientConfigServiceBootstrapConfiguration
```

配置客户端应用在引入 spring-cloud-starter-config 依赖后，其配置的 Bean 都会在 SpringApplication 启动

前加入它的上下文。下面分别介绍这两种配置方式。

■ 通过 HTTP URI 指定 Config Server

通过 HTTP URI 指定 Config Server，是每一个客户端应用默认的启动方式，当 ConfigClient 启动时，通过 spring.cloud.config.uri（如果不配置，默认为 http://localhost:8888）属性绑定到 Config Server，并利用获取到的远端环境属性初始化 Spring 的环境。客户端应用想要获取 Config Server 中的配置信息，需要在环境变量中配置 spring.cloud.config.uri 的值。

ConfigServiceBootstrapConfiguration 进行了两个 Bean 的初始化：ConfigClientProperties 和 ConfigServicePropertySourceLocator。ConfigClientProperties 是对 ConfigClient 的属性进行配置，而 ConfigServicePropertySourceLocator 则用于从远程服务器上请求获取对应服务的配置，并将其注册到 Spring 容器的 Environment 对象中。ConfigClientProperties 的定义如下所示。

```java
public class ConfigClientProperties {
    public static final String PREFIX = "spring.cloud.config";
    public static final String TOKEN_HEADER = "X-Config-Token";
    public static final String STATE_HEADER = "X-Config-State";
    public static final String AUTHORIZATION = "authorization";
    private boolean enabled = true;
    private String profile = "default";
    @Value("${spring.application.name:application}")
    private String name;
    private String label;
    private String username;
    private String password;
    private String[] uri = new String[]{"http://localhost:8888"};
    ......
}
```

从上述代码可以看到，ConfigClientProperties 中定义了 Profile、应用名称、标签、远端服务器的地址等属性。这些都是客户端启动时必需的信息，如果没有这些配置，客户端将不能正确地从 Config Server 获取其对应的配置信息。

另一个属性资源的定位器类 ConfigServicePropertySourceLocator 依赖于客户端应用配置的属性信息，从远程服务器上请求该应用的配置。

下面具体看一下该实现类，如下所示。

```java
// ConfigServicePropertySourceLocator.java
@Retryable(
    interceptor = "configServerRetryInterceptor"
)
public PropertySource<?> locate(org.springframework.core.env.Environment environment) {
......
    if (StringUtils.hasText(properties.getLabel())) {
// 将传入的 label 转换成 labels 数组，label 的格式诸如 dev、test
        labels = StringUtils.commaDelimitedListToStringArray(properties.getLabel());
    }
// 保存请求头部的 X-Config-State
```

```
            String state = ConfigClientStateHolder.getState();
            String[] var9 = labels;
            int var10 = labels.length;

            for(int var11 = 0; var11 < var10; ++var11) {
                String label = var9[var11];
                 org.springframework.cloud.config.environment.Environment result = this.getRemoteEnviron
ment(restTemplate, properties, label.trim(), state);
                if (result != null) {
                    this.log(result);
                    if (result.getPropertySources() != null) {
                        Iterator var14 = result.getPropertySources().iterator();

                        while(var14.hasNext()) {
                                org.springframework.cloud.config.environment.PropertySource source = (org.
springframework.cloud.config.environment.PropertySource)var14.next();
                            Map<String, Object> map = source.getSource();
                            composite.addPropertySource(new MapPropertySource(source.getName(), map));
                        }
                    }
// 其他信息的设置，包括客户端的状态以及版本号等
                    if (StringUtils.hasText(result.getState()) || StringUtils.hasText(result.getVersion())) {
                        HashMap<String, Object> map = new HashMap();
                        this.putValue(map, "config.client.state", result.getState());
                        this.putValue(map, "config.client.version", result.getVersion());
                        composite.addFirstPropertySource(new MapPropertySource("configClient", map));
                    }

                    return composite;
                }
            }
            ......
}
```

重试注解指定了拦截器配置 ConfigServerRetryInterceptor。ConfigServicePropertySourceLocator 实质是一个属性资源定位器，其主要方法是 locate(Environment environment)。首先用当前运行环境的应用名称、Profile 和标签替换 ConfigClientProperties 中的占位符并初始化 RestTemplate，然后遍历 labels 数组直到获取到有效的配置信息，最后根据是否快速失败进行重试。

属性资源定位时调用 GetRemoteEnvironment 方法，通过 HTTP 的方式获取远程服务器上的配置信息。实现过程为，首先替换请求路径中的占位符，然后进行头部组装，组装好了就可以发送请求，最后返回结果。

在上面的实现中，获取到的配置信息存放在 CompositePropertySource 中，那么是如何使用它的呢？这里补充另一个重要的类 PropertySourceBootstrapConfiguration，它实现了 ApplicationContextInitializer 接口，该接口会在应用上下文刷新之前回调 refresh，从而执行初始化操作。应用启动后的调用如下所示。

```
SpringApplicationBuilder.run() -> SpringApplication.run() -> SpringApplication.
createAndRefreshContext() -> SpringApplication.applyinitializers() -> PropertySourceBootstrapConfi
guration.initialize()
```

上述 ConfigServicePropertySourceLocator#locate 方法会在 PropertySourceBootstrapConfiguration#initialize 中被调用，从而保证上下文在刷新之前能够拿到必要的配置信息。initialize 方法如下所示。

```java
// PropertySourceBootstrapConfiguration.java
public void initialize(ConfigurableApplicationContext applicationContext) {
    CompositePropertySource composite = new CompositePropertySource("bootstrapProperties");
    AnnotationAwareOrderComparator.sort(this.propertySourceLocators);
    boolean empty = true;
    ConfigurableEnvironment environment = applicationContext.getEnvironment();
    Iterator var5 = this.propertySourceLocators.iterator();

    while(var5.hasNext()) {
        PropertySourceLocator locator = (PropertySourceLocator)var5.next();
        PropertySource<?> source = null;
        source = locator.locate(environment);
        if (source != null) {
            logger.info("Located property source: " + source);
            composite.addPropertySource(source);
            empty = false;
        }
    }

    if (!empty) {
        MutablePropertySources propertySources = environment.getPropertySources();
        String logConfig = environment.resolvePlaceholders("${logging.config:}");
        LogFile logFile = LogFile.get(environment);
        if (propertySources.contains("bootstrapProperties")) {
            propertySources.remove("bootstrapProperties");
        }

        this.insertPropertySources(propertySources, composite);
        this.reinitializeLoggingSystem(environment, logConfig, logFile);
        this.setLogLevels(applicationContext, environment);
        this.handleIncludedProfiles(environment);
    }
}
```

在 initialize 方法中进行了如下的操作。

（1）根据默认的 AnnotationAwareOrderComparator 排序规则对 propertySourceLocators 数组进行排序。

（2）获取运行环境的上下文 ConfigurableEnvironment。

（3）遍历 propertySourceLocators 时，调用 locate 方法，传入获取的上下文 environment。将 source 添加到 PropertySource 的链表中。设置 source 是否为空的标识变量 empty。

（4）source 不为空，才会设置到 environment 中。

（5）返回 environment 的可变形式，可进行的操作有 AddFirst 和 AddLast。

（6）判断 propertySources 中是否包含 bootstrapProperties，如果包含则移出。

（7）根据 Config Server 覆写的规则，设置 propertySources。

（8）处理多个活跃（Active）Profile 的配置信息。

通过如上过程，可实现将指定的 Config Server 拉取配置信息应用到客户端服务中。

■ 通过服务发现指定 Config Server

Config Client 在启动时，首先会通过服务发现找到 Config Server，然后从 Config Server 拉取其相应的配置信息，并用这些远端的属性资源初始化 Spring 的环境。

如果启用了服务发现，如 Eureka、Consul，则需要设置 spring.cloud.config.discovery.enabled=true，这是因为默认采用 HTTP URI 的方式，会导致客户端应用不能利用服务注册。

所有的客户端应用需要配置正确的服务发现信息。比如使用 Spring Cloud Netflix，使用者需要指定 Eureka 服务器的地址 eureka.client.serviceUrl.defaultZone。在启动时定位服务注册，这样做的开销是需要额外的网络请求，而优点是 Config Server 能够实现高可用，避免单点故障。配置的 Config Server 的 ServiceId 默认是"configserver"，可以通过在客户端的 spring.cloud.config.discovery.serviceId 属性来更改。服务发现的客户端实现支持多种类型的元数据 Map，如 Eureka 的 eureka.instance.metadataMap。Config Server 的一些额外属性需要配置在服务注册的元数据中，这样客户端才能正确连接。如果 Config Server 使用了基本的 HTTP 安全，则可以配置证书的用户名和密码，或者是 Config Server 有一个上下文路径，就可以设置 configPath。客户端中的 Config 信息可以配置如下。

```yaml
eureka:
  instance:
    ......
    metadataMap
      user: admin
      password: 123456
      configPath: /config
```

DiscoveryClientConfigServiceBootstrapConfiguration 中主要配置了 Config Client 通过服务发现组件寻找 Config Server 服务。除此之外还配置了两种事件的监听器，即环境上下文刷新事件和心跳事件，如下所示。

```java
// DiscoveryClientConfigServiceBootstrapConfiguration.java
public void startup(ContextRefreshedEvent event) {
    this.refresh();
}

public void heartbeat(HeartbeatEvent event) {
    if (this.monitor.update(event.getValue())) {
        this.refresh();
    }
}
```

在 Config Client 获取到 Config Server 中的配置信息之后，剩余的过程与指定 HTTP URI 方式获取 Config Server 是一样的。下面将会具体介绍获取配置服务器和事件监听器。

1. 获取配置服务器

ConfigServerInstanceProvider 用于获取配置服务器的地址，对其实例化，会用到服务发现的客户端 DiscoveryClient。其提供的主要方法 getConfigServerInstances，可以通过传入的 ServiceId 参数，获取对应的服务实例，如下所示。

```java
public class ConfigServerInstanceProvider {
    private static Log logger = LogFactory.getLog(ConfigServerInstanceProvider.class);
    private final DiscoveryClient client;
```

```java
    public ConfigServerInstanceProvider(DiscoveryClient client) {
        this.client = client;
    }

    @Retryable(
        interceptor = "configServerRetryInterceptor"
    )
    public List<ServiceInstance> getConfigServerInstances(String serviceId) {
        logger.debug("Locating configserver (" + serviceId + ") via discovery");
        List<ServiceInstance> instances = this.client.getInstances(serviceId);
        if (instances.isEmpty()) {
            throw new IllegalStateException("No instances found of configserver (" + serviceId + ")");
        } else {
            logger.debug("Located configserver (" + serviceId + ") via discovery. No of instances found: " + instances.size());
            return instances;
        }
    }
}
```

上述代码很清晰，主要依赖前面的 DiscoveryClientConfigServiceBootstrap 和 Configuration 注入的对象 DiscoveryClient，通过 client 获取对应 serviceId 的实例。

2. 事件监听器

下面看一下涉及的两种事件，环境上下文刷新事件和心跳事件。

环境上下文刷新事件 ContextRefreshedEvent 的父类继承自抽象类 Application Context Event，当 ApplicationContext 被初始化或者刷新时会唤起该事件，如下所示。

```java
public class ContextRefreshedEvent extends ApplicationContextEvent {
// 创建了一个新的环境上下文刷新事件，参数是初始化了的 ApplicationContext
    public ContextRefreshedEvent(ApplicationContext source) {
        super(source);
    }
}
```

心跳事件定义在 Discovery Client 中，如果支持来自 Discovery Server 的心跳，则在 Discovery Client 的实现中进行广播，并给监听器提供一个基本的服务目录状态变更的指示。目录更新后，该状态值也需要更新，如同一个版本计数器一样，心跳事件的代码如下所示。

```java
public class HeartbeatEvent extends ApplicationEvent {
    private final Object state;

    public HeartbeatEvent(Object source, Object state) {
// 创建一个新的心跳事件，参数通常为 Discovery Client 和状态值
        super(source);
        this.state = state;
    }
// 代表服务目录的状态值
    public Object getValue() {
        return this.state;
    }
}
```

介绍完这两个事件，发现其监听器都依赖于 refresh 方法。下面具体了解一下 refresh 方法的功能，如下所示。

```java
private void refresh() {
    try {
// 获取 ServiceId 的服务实例
        String serviceId = this.config.getDiscovery().getServiceId();
        List<String> listOfUrls = new ArrayList();
        List<ServiceInstance> serviceInstances = this.instanceProvider.getConfigServerInstances(serviceId);

        for(int i = 0; i < serviceInstances.size(); ++i) {
            ServiceInstance server = (ServiceInstance)serviceInstances.get(i);
            String url = this.getHomePage(server);
            String path;
            if (server.getMetadata().containsKey("password")) {
// 刷新获取到的服务实例的元数据信息
                path = (String)server.getMetadata().get("user");
                path = path == null ? "user" : path;
                this.config.setUsername(path);
                String password = (String)server.getMetadata().get("password");
                this.config.setPassword(password);
            }

            if (server.getMetadata().containsKey("configPath")) {
// 更新 configPath
                path = (String)server.getMetadata().get("configPath");
                if (url.endsWith("/") && path.startsWith("/")) {
                    url = url.substring(0, url.length() - 1);
                }

                url = url + path;
            }

            listOfUrls.add(url);
        }

        String[] uri = new String[listOfUrls.size()];
        uri = (String[])listOfUrls.toArray(uri);
        this.config.setUri(uri);
    } catch (Exception var9) {
        if (this.config.isFailFast()) {
            throw var9;
        }

        logger.warn("Could not locate configserver via discovery", var9);
    }

}
```

refresh 方法根据环境上下文刷新事件和心跳事件，刷新服务实例 ConfigClientProperties 中的元数据信息，包括配置的用户名、密码和 configPath。